AMS SHORT COURSE LECTURE NOTES
Introductory Survey Lectures
A Subseries of Proceedings of Symposia in Applied Mathematics

Volume 43 **COMBINATORIAL GAMES**
Edited by Richard K. Guy (Columbus, Ohio, August 1990)

Volume 42 **CRYPTOLOGY AND COMPUTATIONAL NUMBER THEORY**
Edited by C. Pomerance (Boulder, Colorado, August 1989)

Volume 41 **ROBOTICS**
Edited by R. W. Brockett (Louisville, Kentucky, January 1990)

Volume 40 **MATRIX THEORY AND APPLICATIONS**
Edited by Charles R. Johnson (Phoenix, Arizona, January 1989)

Volume 39 **CHAOS AND FRACTALS: THE MATHEMATICS BEHIND THE COMPUTER GRAPHICS**
Edited by Robert L. Devaney and Linda Keen (Providence, Rhode Island, August 1988)

Volume 38 **COMPUTATIONAL COMPLEXITY THEORY**
Edited by Juris Hartmanis (Atlanta, Georgia, January 1988)

Volume 37 **MOMENTS IN MATHEMATICS**
Edited by Henry J. Landau (San Antonio, Texas, January 1987)

Volume 36 **APPROXIMATION THEORY**
Edited by Carl de Boor (New Orleans, Louisiana, January 1986)

Volume 35 **ACTUARIAL MATHEMATICS**
Edited by Harry H. Panjer (Laramie, Wyoming, August 1985)

Volume 34 **MATHEMATICS OF INFORMATION PROCESSING**
Edited by Michael Anshel and William Gewirtz (Louisville, Kentucky, January 1984)

Volume 33 **FAIR ALLOCATION**
Edited by H. Peyton Young (Anaheim, California, January 1985)

Volume 32 **ENVIRONMENTAL AND NATURAL RESOURCE MATHEMATICS**
Edited by R. W. McKelvey (Eugene, Oregon, August 1984)

Volume 31 **COMPUTER COMMUNICATIONS**
Edited by B. Gopinath (Denver, Colorado, January 1983)

Volume 30 **POPULATION BIOLOGY**
Edited by Simon A. Levin (Albany, New York, August 1983)

Volume 29 **APPLIED CRYPTOLOGY, CRYPTOGRAPHIC PROTOCOLS, AND COMPUTER SECURITY MODELS**
By R. A. DeMillo, G. I. Davida, D. P. Dobkin, M. A. Harrison, and R. J. Lipton (San Francisco, California, January 1981)

Volume 28 **STATISTICAL DATA ANALYSIS**
Edited by R. Gnanadesikan (Toronto, Ontario, August 1982)

Volume 27 **COMPUTED TOMOGRAPHY**
Edited by L. A. Shepp (Cincinnati, Ohio, January 1982)

Volume 26 **THE MATHEMATICS OF NETWORKS**
Edited by S. A. Burr (Pittsburgh, Pennsylvania, August 1981)

Volume 25 **OPERATIONS RESEARCH: MATHEMATICS AND MODELS**
Edited by S. I. Gass (Duluth, Minnesota, August 1979)

Volume 24 **GAME THEORY AND ITS APPLICATIONS**
Edited by W. F. Lucas (Biloxi, Mississippi, January 1

PROCEEDINGS OF SYMPOSIA IN APPLIED MATHEMATICS

Volume 23 MODERN STATISTICS: METHODS AND APPLICATIONS
Edited by R. V. Hogg (San Antonio, Texas, January 1980)

Volume 22 NUMERICAL ANALYSIS
Edited by G. H. Golub and J. Oliger (Atlanta, Georgia, January 1978)

Volume 21 MATHEMATICAL ASPECTS OF PRODUCTION AND DISTRIBUTION OF ENERGY
Edited by P. D. Lax (San Antonio, Texas, January 1976)

Volume 20 THE INFLUENCE OF COMPUTING ON MATHEMATICAL RESEARCH AND EDUCATION
Edited by J. P LaSalle (University of Montana, August 1973)

Volume 19 MATHEMATICAL ASPECTS OF COMPUTER SCIENCE
Edited by J. T. Schwartz (New York City, April 1966)

Volume 18 MAGNETO-FLUID AND PLASMA DYNAMICS
Edited by H. Grad (New York City, April 1965)

Volume 17 APPLICATIONS OF NONLINEAR PARTIAL DIFFERENTIAL EQUATIONS IN MATHEMATICAL PHYSICS
Edited by R. Finn (New York City, April 1964)

Volume 16 STOCHASTIC PROCESSES IN MATHEMATICAL PHYSICS AND ENGINEERING
Edited by R. Bellman (New York City, April 1963)

Volume 15 EXPERIMENTAL ARITHMETIC, HIGH SPEED COMPUTING, AND MATHEMATICS
Edited by N. C. Metropolis, A. H. Taub, J. Todd, and C. B. Tompkins (Atlantic City and Chicago, April 1962)

Volume 14 MATHEMATICAL PROBLEMS IN THE BIOLOGICAL SCIENCES
Edited by R. Bellman (New York City, April 1961)

Volume 13 HYDRODYNAMIC INSTABILITY
Edited by R. Bellman, G. Birkhoff and C. C. Lin (New York City, April 1960)

Volume 12 STRUCTURE OF LANGUAGE AND ITS MATHEMATICAL ASPECTS
Edited by R. Jakobson (New York City, April 1960)

Volume 11 NUCLEAR REACTOR THEORY
Edited by G. Birkhoff and E. P. Wigner (New York City, April 1959)

Volume 10 COMBINATORIAL ANALYSIS
Edited by R. Bellman and M. Hall, Jr. (New York University, April 1957)

Volume 9 ORBIT THEORY
Edited by G. Birkhoff and R. E. Langer (Columbia University, April 1958)

Volume 8 CALCULUS OF VARIATIONS AND ITS APPLICATIONS
Edited by L. M. Graves (University of Chicago, April 1956)

Volume 7 APPLIED PROBABILITY
Edited by L. A. MacColl (Polytechnic Institute of Brooklyn, April 1955)

Volume 6 NUMERICAL ANALYSIS
Edited by J. H. Curtiss (Santa Monica City College, August 1953)

Volume 5 WAVE MOTION AND VIBRATION THEORY
Edited by A. E. Heins (Carnegie Institute of Technology, June 1952)

Volume 4 FLUID DYNAMICS
Edited by M. H. Martin (University of Maryland, June 1951)

PROCEEDINGS OF SYMPOSIA IN APPLIED MATHEMATICS

Volume 3 **ELASTICITY**
Edited by R. V. Churchill (*University of Michigan, June 1949*)

Volume 2 **ELECTROMAGNETIC THEORY**
Edited by A. H. Taub (*Massachusetts Institute of Technology, July 1948*)

Volume 1 **NON-LINEAR PROBLEMS IN MECHANICS OF CONTINUA**
Edited by E. Reissner (*Brown University, August 1947*)

AMS SHORT COURSE LECTURE NOTES

Introductory Survey Lectures

published as a subseries of

Proceedings of Symposia in Applied Mathematics

PROCEEDINGS OF SYMPOSIA
IN APPLIED MATHEMATICS
Volume 43

Combinatorial Games

Richard K. Guy, Editor

Elwyn R. Berlekamp
John Horton Conway
Aviezri S. Fraenkel
Richard J. Nowakowski
Vera Pless

American Mathematical Society
Providence, Rhode Island

LECTURE NOTES PREPARED FOR THE
AMERICAN MATHEMATICAL SOCIETY SHORT COURSE

COMBINATORIAL GAMES

HELD IN COLUMBUS, OHIO
AUGUST 6–7, 1990

The AMS Short Course Series is sponsored by the Society's Committee on Employment and Educational Policy (CEEP). The series is under the direction of the Short Course Advisory Subcommittee of CEEP.

2000 *Mathematics Subject Classification.* Primary 91A05; Secondary 94B99.

Library of Congress Cataloging-in-Publication Data

Combinatorial games/Richard K. Guy, editor.
p. cm. — (Proceedings of symposia in applied mathematics; v. 43. AMS short course lecture notes)
ISBN 0-8218-0166-X
1. Game theory—Congresses. 2. Combinatorial analysis—Congresses. I. Guy, Richard K. II. Series: Proceedings of symposia in applied mathematics; v. 43. III. Series: Proceedings of symposia in applied mathematics. AMS Short course lecture notes.
QA269.C65 1991 90-22771
519.3—dc20 CIP

∞ The paper used in this book is acid-free and falls within the guidelines
established to ensure permanence and durability.
Printed in the United States of America.

This publication was prepared by the authors using TEX.
Visit the AMS home page at URL: http://www.ams.org/

10 9 8 7 6 5 4 3 04 03 02 01 00

Table of Contents

Preface

The subject of combinatorics is only slowly acquiring respectability and combinatorial games will clearly take longer than the rest of combinatorics. Perhaps this partly stems from the puritanical view that anything amusing can't possibly involve any worthwhile mathematics.

In the past, "game theory" has meant the subject delineated by von Neumann and Morgenstern, which has found wide, though usually unsuccessful, application in economics, management, military strategy, and other useful forms of human activity. Combinatorial games, with complete information, no chance moves, and no place for bluffing or coalitions, are of little interest to the classical game theorist, who knows that there is always at least one pure optimal strategy.

Why, then, should we be interested in combinatorial games?

Aviezri Fraenkel gives some cogent reasons in the introduction to his Selected Bibliography at the end of this volume.

There are many connexions with other parts of mathematics, only a few of which have so far begun to be explored, and only one of which is seriously considered here. Vera Pless explains the several connexions with coding theory, through which we make contact with most of the branches of the main stream of combinatorics, including graph theory.

A whole new theory of number, including infinitesimals and transfinite numbers, has emerged as a special case of the theory of games. This is introduced by John Conway in the second chapter. Investigation of this remarkable area is only slowly gaining momentum, in spite of the early appearance of Donald Knuth's popularization, *Surreal Numbers.* Perhaps this only served to perpetuate the myth that we are dealing with a frivolous subject.

As Aviezri Fraenkel explains, complexity theory is very well illustrated by combinatorial games, which supply a plethora of examples of harder problems than most of those which have been considered in the past.

We have been able to do no more than touch on the theory of "hot" games, which are, of course, the interesting ones from a practical, as well as a theoretical point of view. Elwyn Berlekamp explains the significant progress that he has been making with the analysis of endgames in Go, a game long thought to be even more intractable than Chess.

We introduce "impartial" games in Chapter 3. These "tepid" games are of minimal interest to the classical game theorist, but there are plenty of unsolved problems in this area, as well as in the rest of the subject. A notable one is how to deal with the "misère play" of impartial games. A small but important break-through was recently made by William Sibert, and Thane Plambeck is now unveiling the misère analysis of several games which had earlier seemed intransigent.

A list of open problems is given towards the end of the book. While some of these are undoubtedly hard, inroads are being made into others even as I write, and a new generation of graduate students will find a rich vein of material waiting to be investigated.

As examples of what has been and what remains to be done, Richard Nowakowski presents in the last chapter some specific examples of games. Welter's Game is now understood, but see the quotation of Berlekamp by Fraenkel in section 6.2 of the "complexity" chapter. In Conway's game of Sylver Coinage, on the other hand, much remains to be discovered, though G.L. Sicherman and others are slowly revealing more and more of the truth. Finally, Berlekamp's masterly analysis of the well-known children's game of Dots-and-Boxes illustrates the many levels at which a seemingly simple game may be played, and the many quite sophisticated techniques which may be used in its analysis.

We are indebted to Academic Press for permission to reproduce some text and a number of figures from John Conway's *On Numbers and Games* and from *Winning Ways for your Mathematical Plays* by Berlekamp, Conway and Guy.

Thanks to the Short Course Committee for suggesting and to Jim Maxwell and Monica Foulkes for organizing the course that has supplied the *raison d'être* for the present volume, and to Alison Buckser and other members of the American Mathematical Society's staff for its careful and expert production.

Richard K. Guy,
The University of Calgary,
90-11-05.

Proceedings of Symposia in Applied Mathematics
Volume **43**, 1991

What is a Game?

Richard K. Guy

1 Introduction

We only skim the surface of the vast subject of combinatorial games. For more breadth, depth and detail, consult both of the books:

Elwyn R. Berlekamp, John H. Conway and Richard K. Guy, Winning Ways for your Mathematical Plays, Academic Press, London & New York, 1982;

John H. Conway, On Numbers and Games, *London Math. Soc. Monograph* **6**, Academic Press, London & New York, 1956;

which we will frequently refer to as WW and ONAG respectively.

Two other surveys are Fraenkel (1980), who considers the complexity of games, and Guy (1983), who explores the connexions between games and graphs.

Fraenkel contrasts Nim with Go, the former with a very simple winning strategy, the latter very complicated. In later lectures in this course, Fraenkel will discuss complexity, I will go into some detail about impartial games, whose prototype is the game of Nim, and Berlekamp will tell you about his discoveries concerning the game of Go. Nim has no cycles in its game graph, no interaction between tokens, and is impartial; Go has cycles and interaction and is partizan. The spectrum between the two games spans the complexity gap between polynomial, Pspace-complete, and Extime-complete games. In existential problems such as the travelling salesman problem, high complexity is a liability, but in games and cryptanalysis, it can be an asset.

The paper is in final form and no version of it will be submitted for publication elsewhere.
1980 *Mathematics Subject Classification* (1985 *Revision*). Primary 90D05.

Fraenkel also maintains a valuable bibliography of the subject, copies of which may be obtained from him at the Weizmann Institute, Rehovot, Israel. A recent edition appears at the end of the present volume.

Guy surveys the connexions between combinatorial game theory and graph theory: graphs of games; games on graphs (Hackenbush, von Neumann's game, Rims, Rails, Lucasta, Sprouts); the ways graphs can be used to elucidate puzzles (Tantalizer, Rubik's Cube, Fifteen Puzzle, magic squares); and the occurrence of Euler's formula in Berlekamp's analysis of Dots-and-Boxes [WW, 507–550].

2 What We Mean by a Combinatorial Game

[WW, 16–17]

Our games are unlike those of "classical" game theory, that find application in economics, management, and military strategy. Our games (almost always!) satisfy the following conditions:

1. There are just two players, often called Left and Right. There can be no question of **coalitions**.
2. There are several, usually finitely many, **positions**, and often a particular **starting position**.
3. There are clearly defined **rules** that specify the two sets of **moves** that Left and Right can make from a given position to its **options**.
4. Left and Right move alternately, in the game as a whole.
5. In the **normal play** convention a player unable to move **loses**.
6. The rules are such that play will always come to an end because some player will be unable to move. This is called the **ending condition**. There are no games which are drawn by repetition of moves.
7. Both players know what is going on; there is **complete information**. There is no occasion for **bluffing**.
8. There are no **chance moves**: no dealing of cards; no rolling of dice.

Think about games that you know. How far do they satisfy these eight conditions?

Ludo, **Snakes-and-Ladders** and **Backgammon** all have complete information, but contain chance moves, since they all use dice.

Battleships, **Kriegspiel**, **Three-Finger Morra** and **Scissors-Paper-Stone** have no chance moves but the players do not have complete information about the disposition of their opponents' pieces or fingers. Moreover, in the finger games, the players move simultaneously rather than alternately.

Tic-Tac-Toe (**Noughts-and-Crosses**) fails 5. because a player unable to move is not necessarily the loser, since ties are possible. **Chess** also fails 5. and contains positions that are *tied* by stalemate (in which the last player does *not* win) and positions that are *drawn* by infinite play (of which perpetual check is a special case). The words "tied" and "drawn" are often used interchangeably, though with slight transatlantic differences, for games which are neither won nor lost. We suggest that **drawn** be used for cases when this happens because play is drawn out indefinitely and **tied** for cases when play definitely ends but the rules do not award a win to either player.

Monopoly fails on several counts. As in Ludo, there are chance moves and there may be more than two players. The players don't have complete information about the arrangement of the cards and the game could, theoretically, go on for ever.

In **Poker** much of the interest derives from the incompleteness of the information, the chance moves and the possibility of coalitions.

Bridge is a two-person game, each "person" being a team of two, but the players do not even have complete information about their own cards.

Nim is played with heaps of beans. When it is your turn to move, choose a heap and remove as many beans from it as you wish; perhaps the whole heap, but at least one bean. **Grundy's Game** is also played with heaps of beans. A move now is to split a heap into two heaps of *unequal* size, so that heaps of one or two cannot be split. **Wythoff's Game** is played with just two heaps. A move is to remove any number of beans from one heap, or *equal* numbers of beans from both heaps. This last option is an example of a **nondisjunctive** move: it doesn't satisfy the condition for the **sum** of two games, which we will define later. Nim, Grundy's Game, and Wythoff's Game each satisfy all of our eight conditions, together with the additional one that the options from any given position are the same for each player, regardless of whose turn it is to move. Such games are called **impartial**; those games in which the options for the two players are not all alike are called **partizan**.

Dots-and-Boxes is won by the player scoring the larger number of boxes, so that it doesn't satisfy the normal play convention. However, as Richard Nowakowski explains in the last chapter, it can almost always be treated as an impartial game, satisfying the normal play convention. Part of its theory uses that of **Kayles** and of **Dawson's Kayles**. These two games are played with rows of skittles: in Kayles a move removes just one skittle, or two adjacent ones, so that the game repeatedly splits into a sum of smaller games. Similarly in Dawson's Kayles, in which the move is to remove any skittle, provided that its immediate neighbors, if any, are also removed.

Sylver Coinage is an impartial game which uses the **misère play convention** that the last person to play *loses*. Misère games are usually very difficult to analyze, though a recent breakthrough was made by William Sibert and John Conway who have found the complete analysis of **Misère Kayles**.

Go is a good example of a "hot" partizan game, i.e. one in which, in the great majority of positions, each player is eager to make the next move. As Elwyn Berlekamp explains in a later chapter, he has recently been making good progress with analyzing the concluding stages of the game, using his generalization of the idea of "overheating" [WW 170–174].

We will often refer to a move as being "good" if it wins, and "bad" if it won't. In theory it usually suffices to find any good move, or to show than no good move exists. But in real life games there are many other criteria for choosing between your various options. If you're *losing*, then all your options are bad in the above sense, but in practice they're not all equal, and you might prefer one that makes the situation too complicated for your opponent to analyze (the **Enough Rope Principle**).

It is hard to draw the line between mathematics and psychology. There are even cases where you should prefer a bad move to a good one! Your opponent might be learning how to play a game with which you're already familiar. In this case you'll probably be able to win a few times despite the bad moves you deliberately make so as not to give away your strategy. Or one move, theoretically the best, might gain you only a dollar, while another, which loses a dollar, might win you a hundred if your opponent fails to find the subtle winning reply. Or you may be a baby-sitter, whose job is much more peaceful if your opponent wins. Or a card-sharp who's losing while the stakes are low, in anticipation of winning later when the stakes are higher.

3 Game Graphs and Trees

A game may be visualized as a digraph: the nodes are the positions and the arcs are the options. The arcs may be thought of as colored, say

bLue,	Red, or	grEen
	according as the option is available	
to Left only,	to Right only, or	to Either player.

Alternatively, we may distinguish between different plays of the game, i.e. different dipaths in the digraph, by duplicating the nodes as necessary and representing the game by a rooted tree. The root is the starting position and the arcs are directed away from the root.

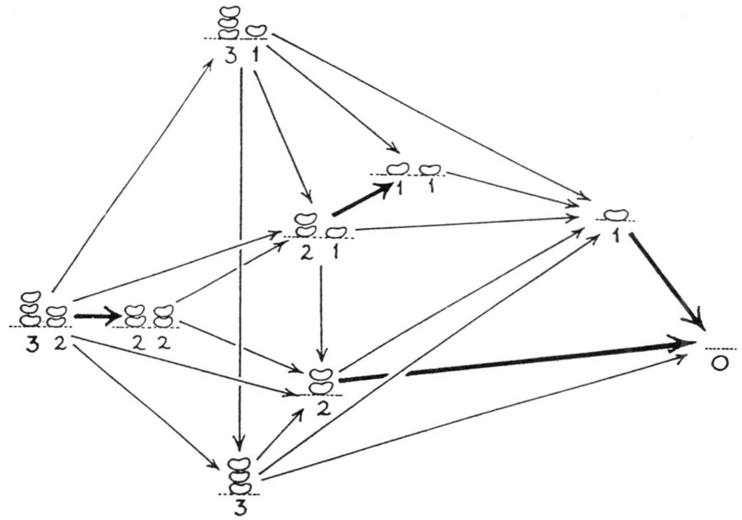

Figure 1: The game graph for the Nim position {3,2}.
From the leftmost position the next player can win
by adopting the strategy indicated by the heavy arrows.

Figures 1 and 2 show the game graph and the game tree for the position {3,2} in a game of Nim: two heaps, one with three beans, the other with two. Nim is an example of an impartial game, in every position of which the same set of options is available to either player: think of the arcs in Figures 1 and 2 as being colored green. Nim is played with a number of heaps of beans. The typical option, for either player, is to choose a heap and remove from it as many beans as you wish: the whole heap maybe, but at least one bean.

Notice the difference between the **complete analysis** of a game, and a **winning strategy**. Figure 2 is a complete analysis: for a winning strategy it suffices to describe the four black arrows.

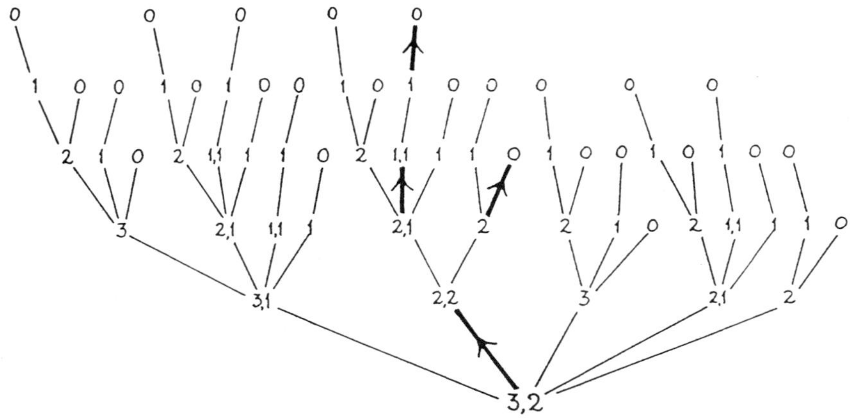

Figure 2: The game tree for the same Nim position.

The root is {3,2}, and the arcs are directed upwards.

You may mentally identify three seemingly different aspects of the same idea.

1. A **game**, i.e. the whole digraph or tree representing the game. For example, **the** Game of Chess, as opposed to **a** (particular) game of chess. We refer to the latter as a **play** of the game: compare *le jeu d'échecs, une partie d'échecs.*

2. A **position** in a game; a particular node of the digraph, perhaps the root of the tree. For example, the standard opening position in Chess, ready for a play of the game.

3. The **ordered pair** of sets of options available to the two players from a given position, e.g.

{Pa3,Pa4,...,Ph3,Ph4,Sa3,Sc3,Sf3,Sh3|Pa6,Pa5,...,Ph6,Ph5,Sa6,Sc6,Sf6,Sh6}

A position, such as the rightmost in Figure 1, or any zero in Figure 2, from which neither player has any option, is a **terminal position**, at which the game ends. The **outcome** is then specified by the rules. It may be a win for Left, or a win for Right, possibly accompanied by some score or payoff. The rules may not specify a winner, so that the game may end in a **tie**. For present purposes we will adopt the **normal play convention** that the winner is the player who has just made the last move: equivalently, **last player winning**; if you can't move, you lose. We won't have time to say very much about the **misère play convention**, which accords the win to a player unable to move: **last player losing**. Analysis is far more difficult in this case.

To ensure that we *have* a last player, our games must end. We assume that they satisfy the **ending condition**: that there is no infinite sequence of options. Notice that this condition prohibits *all* infinite sequences, not merely those in which Left and Right make alternate moves. In order to give **values** to our games, we need to consider the possibility of several consecutive moves by the same player. This can occur in the play of the **sum** of two or more games, as we shall see.

A game that does not satisfy the ending condition is called a **loopy game**. Its digraph will contain a directed circuit or an infinite directed path. The outcome may be a **draw**: note that we distinguish between a *tied* game and one *drawn out* by infinite play. Chess exhibits both kinds of outcome: stalemate is a tie, but perpetual check, repetition of moves, or insufficient mating material are equivalent to draws.

4 The Formal Definition of a Game

This is deceptively simple: each game is an ordered pair of sets of games:

$$G = \{\{G^{L_1}, G^{L_2}, \ldots\}|\{G^{R_1}, G^{R_2}, \ldots\}\}.$$

To avoid proliferation of braces, we write this more compactly as

$$G = \{G^L|G^R\}$$

where we must remember that G^L and G^R are *sets* of Left and Right options, which may, for example, be infinite, or empty. Indeed the definition is inductive, and the empty set is the basis for the induction, which starts with the **Endgame**

$$\{\emptyset|\emptyset\} = \{\ \mid\ \},$$

in which neither player has an option, and which we will denote by 0 (**zero**).

Here, and from now on, we use several symbols, which are familiar in elementary arithmetic, with the strong implication that we can manipulate games in the same way that we manipulate numbers in ordinary arithmetic. Some games behave like numbers and we call them numbers, but to justify the manipulations takes more space than we have here, so turn to pages 71–96 of ONAG if you would like more detail and further examples. See also the next chapter, Numbers and Games, by John Conway.

It's helpful to attach ordinal numbers, or **birthdays**, to games, and to introduce the idea of **simplicity** [WW, 23–27]. When a move is made in a game, it becomes *simpler* in the sense that we arrive at a position with an earlier birthday. All definitions and proofs are inductive in that they are assumed to have been made for all simpler games. The basis, as we have already stated, is the simplest game of all, the Endgame, born on day zero.

On day one we have two sets, the empty set and the set $\{0\}$ consisting of the Endgame; so that we can visualize 2^2 games. Their game trees (in which Left's moves slope up to the left and Right's moves slope up to the right), together with their names, are shown in Fig. 3.

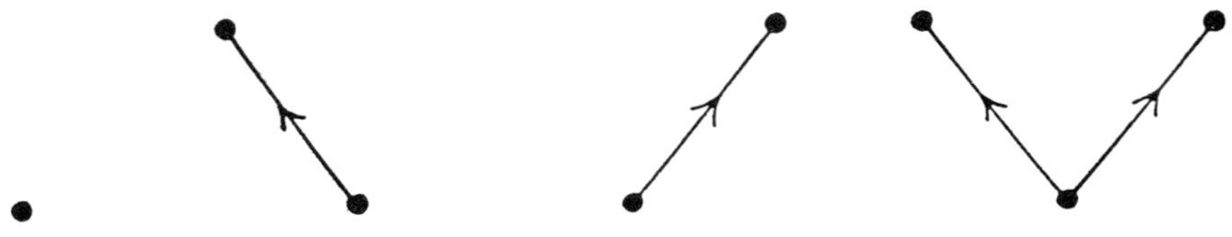

$0 = \{\ |\ \}$ $1 = \{0|\ \}$ $-1 = \{\ |0\}$ $* = \{0|0\}$

Figure 3: The four simplest games, born on days zero and one.

We quote from p. 72 of ONAG:

> The simplest game of all is the *Endgame*, 0. I courteously offer you the first move in this game, and call upon you to make it. You lose, of course, because 0 is defined as the game in which it is never legal to make a move.
>
> In the game $1 = \{0|\}$, there is a legal move for Left, which ends the game, but at no time is there any legal move for Right. If I play Left, and you Right, and you have first move again (only fair, as you lost the previous game) you will lose again, being unable to move even from the initial position. To demonstrate my skill, I shall now start from the same position, make my legal move to 0, and call upon you to make yours.
>
> Of course you are now beginning to suspect that Left always wins, so for our next game, -1, you may play as Left and I as Right! For the last of our examples, the new game $* = \{0|0\}$, you may play whichever role you wish, provided that for this privilege you allow me to play first.

In summary:

The Endgame is the prototype of games in which the next player loses, since no option is available: a **second player win.**

The game 1 is a **Left win**, no matter who starts: if Louise starts, she goes to $\{\ |\ \} = 0$ and Richard has no option and loses; if Richard starts, he has no option and loses even more quickly.

The game -1 is a **Right win**, no matter who starts.

The game $\{0|0\} = *$ ("Star") is the simplest game which is not a number [WW, 40]. It is a **first player win.**

5 The Four Outcome Classes

If we adopt the normal play convention, every game belongs to just one of four outcome classes [ONAG, Theorem 50] which are exemplified by the four games we've just seen. The terminology and notation are displayed in Figure 4.

<table>
<tr><td colspan="2" rowspan="2">If, in a game G</td><td colspan="2">Right starts</td></tr>
<tr><td>& L has a winning strategy</td><td>& R has a winning strategy</td></tr>
<tr><td rowspan="2">Left starts</td><td>& R has a winning strategy</td><td>ZERO
$G = 0$
2nd wins</td><td>NEGATIVE
$G < 0$
R wins</td></tr>
<tr><td>& L has a winning strategy</td><td>POSITIVE
$G > 0$
L wins</td><td>FUZZY
$G \| 0$
1st wins</td></tr>
</table>

Figure 4: The four outcome classes.

It is convenient to combine these outcome classes and symbols in pairs.

If	Left	Right	Left	Right	has a winning strategy
provided	Right	Left	Left	Right	starts, then we
write	$G \geq 0$	$G \leq 0$	$G \Vert\!\triangleright 0$	$G \triangleleft\!\Vert 0$	corresponding to
the	1st col	1st row	2nd row	2nd col	of Figure 4.

6 The Negative of a Game

A device to breathe new life into an otherwise one-sided contest is to allow a novice opponent, when he feels he is losing, to turn the board around, to reverse the roles of the two players, to handicap his more skilled adversary, by asking her to defend what appears to him to be an inferior position. This replaces the game by its negative. Formally, the **negative** of G,

$$-G = \{-G^R | -G^L\}$$

is defined inductively [WW, 35]. Remember that $-G^R$, for example, is short for the set $\{-G^{R_1},\ -G^{R_2},\ \ldots\}$, whose members are simpler games than $-G$, and hence have been defined earlier.

7 Sums of Games

There are many ways of playing two or more games simultaneously, but often the most natural is what we call the **sum**, or **disjunctive compound** [ONAG, 75; WW, 33]. Nim, for example, is the sum of a number of games of one-heap Nim. In the sum of two or more component games, the player whose turn it is to move selects one component and makes a legal move in it:

$$G + H = \{G^L + H,\, G + H^L | G^R + H,\, G + H^R\}.$$

Once again this is an inductive definition: $G^R + H$, for example, represents the set of options $\{G^{R_1} + H,\ G^{R_2} + H,\ \ldots\}$ each of which is a simpler game than $G + H$, so that addition there is already defined.

It's not hard to see that sums are commutative and associative, that $G + 0 = G$, and [ONAG, Theorem 51] that $G + (-G) = 0$. In that last sentence we've used zero in two quite different senses. In $G + 0 = G$ we intended 0 to mean the Endgame, $\{\ |\ \}$. In $G + (-G) = 0$ we intended

"$=0$" to mean "is a zero game", that is "belongs to the (very large!) equivalence class of games for which the second player has a winning strategy". Check that $1+(-1)=0$ and that $*+*=0$, so that we can speak of the games $1+(-1)$ and $*+*$ as having the same **value**, 0, as the Endgame, even though their **forms** are different.

More generally, we will say that two games are **equivalent**, and have the same **value**, and write $G=H$, if the game $G+(-H)$ is a second player win. With the above definitions of sum, negative and zero, the set of all games forms a commutative group. Moreover, games form a partially ordered set, and we write $G>H$ just if $G-H>0$, that is, if Left can win the sum $G+(-H)$, no matter who starts. Our notation is justified by theorems such as the following, proved in [ONAG, 76]. If $G \geq 0$ and $H \geq 0$, then $G+H \geq 0$. If H is a zero game (that is, a win for the second player), then $G+H$ has the same outcome as G. If $H-K$ is a zero game, then $G+H$ and $G+K$ have the same outcome.

8 The Games Born on Day Two

As day two dawns we have four games to play with, and so $2^4=16$ sets of games. There are 16 choices for Left's options and 16 for Right's, giving a potential of 256 games on day two. However, things are not *quite* that complicated, in that for each player, some options are clearly preferable to others. The four games born on day one can be arranged in the lattice (in the poset sense, rather than the geometrical sense) of Figure 5, in which Left's preferences are placed higher, and Right's are lower.

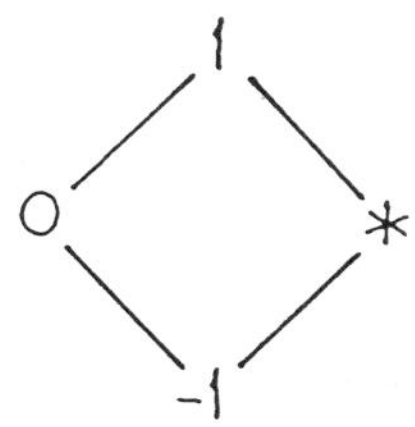

Figure 5: The lattice of games born on day one.

The only set of options for which there is any doubt in either player's mind about the best move, is the incomparable pair $\{0,*\}$. So, for a player's options we need consider only six possibilities: the empty set, the four singletons, and this incomparable pair. Among the resulting 6^2 possibilities for games born on day two, just 22 are inequivalent and

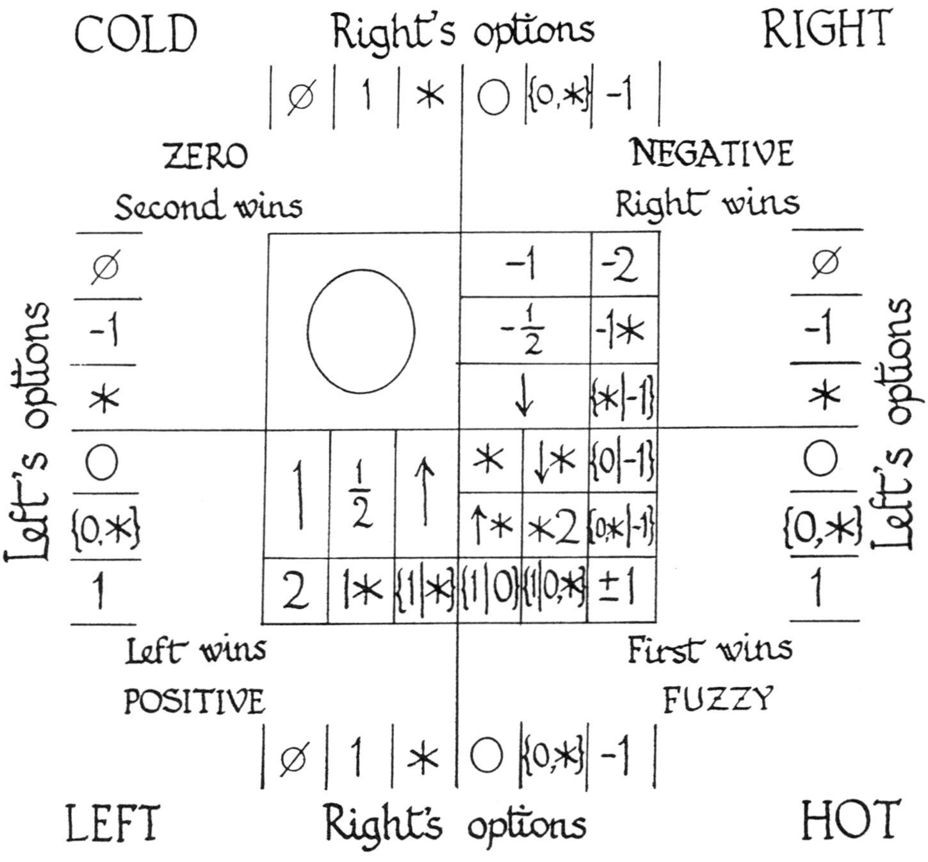

Figure 6: The twenty-two games born on day two.

18 are new. They are shown in Figure 6, the four quarters of which should be compared with those of Figure 4. These contain the zero game; six negative games; six positive; and nine fuzzy ones. The six sets of Right options run from left to right in increasing order of desirability from Right's point of view; Left's run correspondingly downwards.

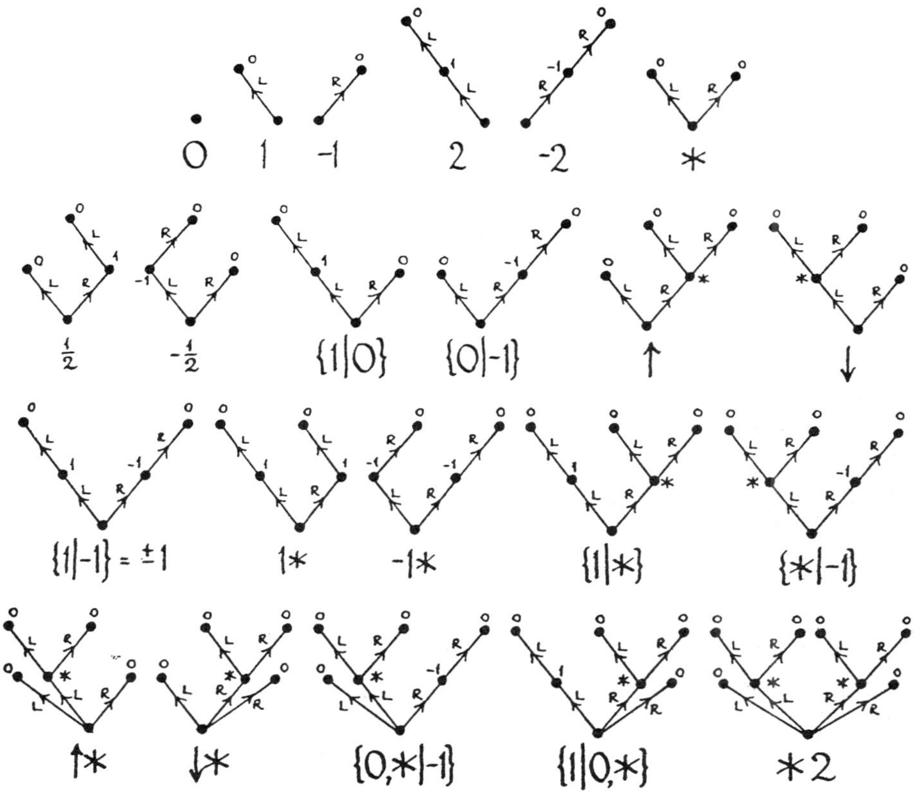

Figure 7: The game trees for the twenty-two games born on day two.

Figure 7 shows the game trees for these twenty-two day-two games: the lowest node in each case is the root, arcs sloping up to the left are blue, those sloping up to the right are red. The trees for Star and sixteen of the day-two games have been condensed into digraphs in Figure 8; arcs labelled E are green and represent options available to both Left and Right.

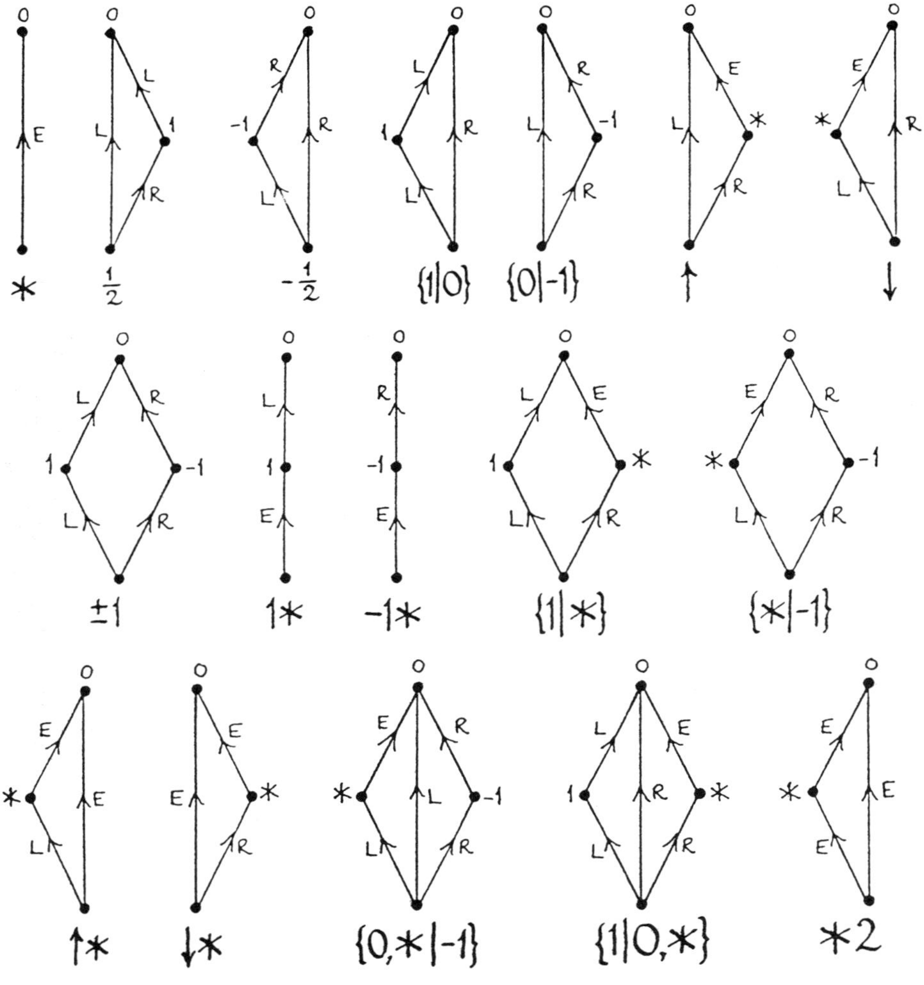

Figure 8: Digraphs for Star and sixteen day-two games.

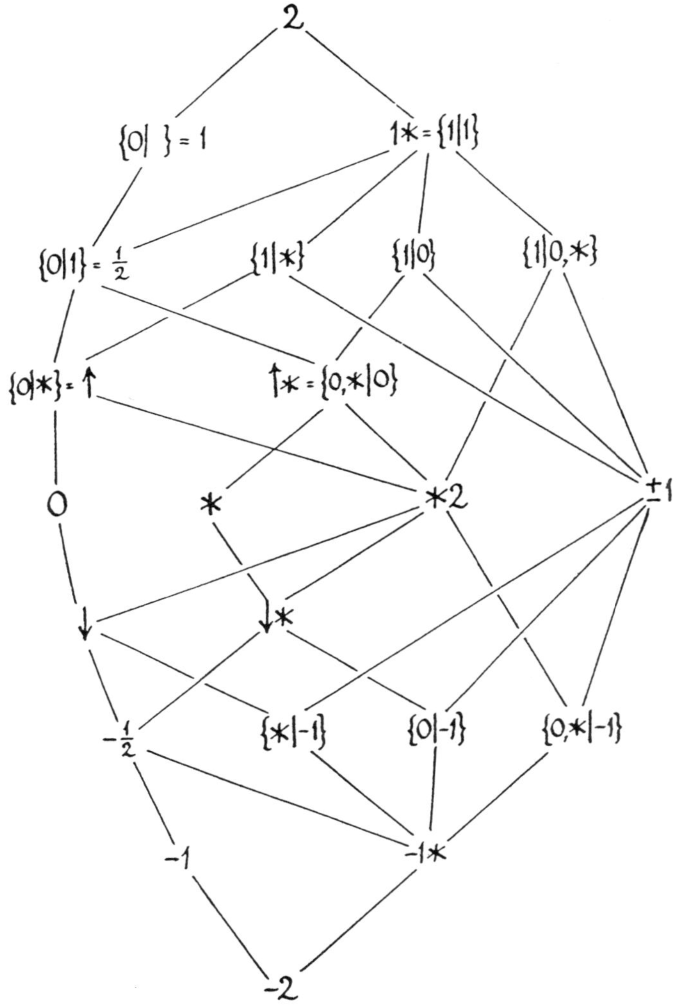

Figure 9: The lattice of day-two games.

The twenty-two games are exhibited as a lattice in Figure 9. If two games are connected by an arc, or, transitively by a path of arcs, then the higher game has a greater value than the lower, as in Figure 5.

Examples and exercises.

$1+1=2$, $\frac{1}{2}+\frac{1}{2}=1$, $*+*=0$, $\uparrow=\{0|*\}=\{0,*|*\}>0$ ("Up is positive"), $\uparrow *=\{0,*|0\}=\uparrow+*$ ("Upstar"), $0 \parallel *2=\{0,*|0,*\}$ ("0 is incomparable with Star-two"), $1*=\{1|1\}=1+*$, $\{1|0\}>*$, $\downarrow *=\{0|0,*\} \parallel 0$ ("Downstar is incomparable with zero"), $\{1|0\}>\uparrow *$, $\{1|0\}>*2$, $\{1|0\}>\downarrow=\{*|0\}$, $\{1|0\}>\downarrow *$, $\{1|0\}>\pm 1=\{1|-1\}$, $\{1|0\} \parallel 0$, $\{1|0\} \parallel \uparrow$, $\{1|*\}>0$, $\{1|*\}>\uparrow$, $\{1|*\} \parallel \uparrow *$, $\{1|0,*\}>\downarrow$, $\{1|0,*\} \parallel \uparrow$.

Two important ways of classifying games are as **hot** or **cold** and as **partizan** or **impartial**. We will shortly make a brief attempt to distinguish hot from cold. Impartial games are those in which the two sets of Left & Right options are the same; in partizan games the two sets are different, in general. As illustrative examples, we describe two partizan games: Domineering is a hot game; Blue-Red Hackenbush is a cold one.

9 Domineering

[ONAG 74–75, 120–121; WW 117–120, 137–140.] This is also called Crosscram. Left and Right alternately place dominoes so that they exactly cover two squares of a checker-board. Left orients her dominoes North-South and Right puts his East-West. Dominoes mustn't overlap each other or the edge of the board. A player loses who can find no appropriately oriented space for a domino. After a while, the available space may separate into disconnected regions, and the game becomes the sum of smaller games. Many of the games we have already seen are realized by small Domineering "boards". Check the values in Figure 10.

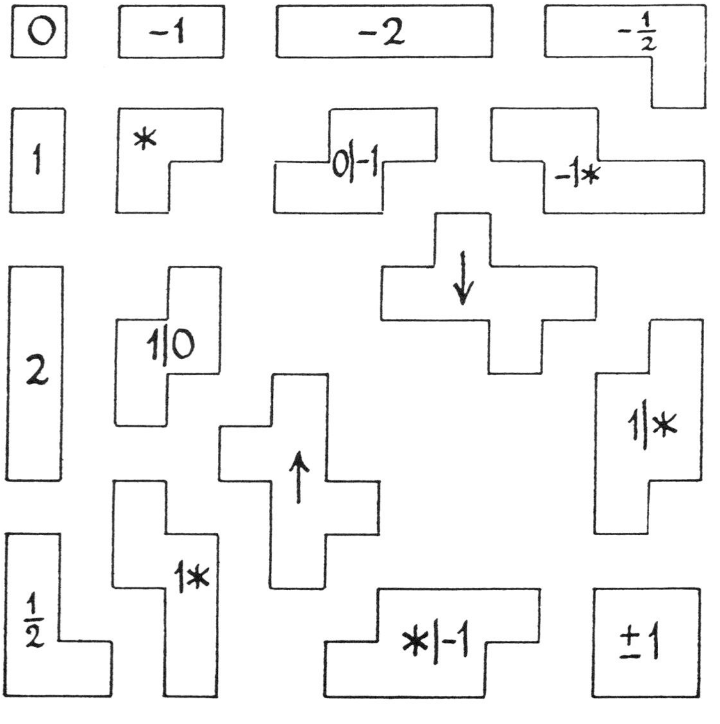

Figure 10: Small Domineering boards realize all four day-one games (top left corner) and thirteen day-two games.

10 Hot and Cold Games

Domineering is an example of a **hot game**. These are the interesting games in which there's an advantage in having the move: the first player wins. If $G = \{G^L|G^R\}$, then the various differences $G^L - G$ and $G - G^R$ are the (Left and Right) **incentives** of G. These are always negative if G is a number. Numbers are **cold games** and the Number Avoidance Theorem tells you that you should

Never move in a Number, unless there's Nothing else to do.

An earnest of the theory of hot games can be found in the work of Milnor (1953) and Hanner (1959). For recent developments, see WW, 141–182 and Berlekamp (1988), who is currently generalizing the theory of "overheating" and making inroads into the difficult theory of the game of Go, some of which are described in his later chapter. See also the last section of the next chapter which describes the thermograph of a game.

A good example of a cold game is

11 Blue–Red Hackenbush

This is perhaps best played on a blackboard, using an eraser. Start with a **picture**, for example Figure 11, which is a graph, some of whose nodes are on the ground (the dotted line), and whose edges are either blue or red [ONAG, 86–91; WW, 3–8]. A Left (resp. Right) move is to delete a blue (resp. red) edge, together with any edges of either color which are no longer connected to the ground.

If Right deletes the dog's neck, for example, the head disappears as well. If Left deletes the body, no other edges disappear, but the picture breaks into the sum of two separate pictures. The aim, as usual, is to be the last player, the person whose move leaves no edges of the opponent's color.

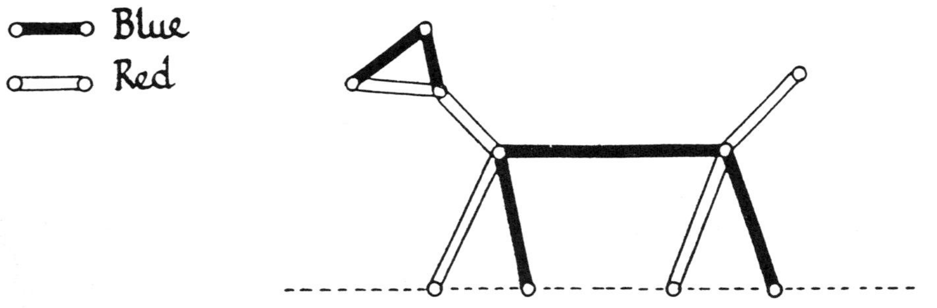

Figure 11: A Blue–Red Hackenbush picture.

12 When is a Game a Number?

Although the values of Blue–Red Hackenbush pictures may be hard to calculate (in the technical sense; WW, 210–212), they are all numbers. A game is a **number** [ONAG, 81] exactly if all its options are numbers and no Left option is greater than or equal to any Right option. The game $\pm 1 = \{1 | -1\}$ is *not* a number, since $1 \geq -1$; Star is not a number, because $0 \geq 0$; and $\uparrow = \{0 | *\}$ is not a number, because $*$ is not. Examples of numbers are

0	born on day 0
1 and -1	born on day 1
$\frac{1}{2}$, $-\frac{1}{2}$, 2 and -2	born on day 2
$\frac{1}{4}$, $-\frac{1}{4}$, $\frac{3}{4}$, $-\frac{3}{4}$, $1\frac{1}{2}$, $-1\frac{1}{2}$, 3 and -3	born on day 3

On day ω *all* the remaining real numbers are born (as Dedekind sections of the binary rationals), as well as the first infinite ordinals, $\omega = \{0,\ 1,\ 2,\ \ldots\ |\ \}$ and $-\omega = \{\ |0,\ -1,\ -2,\ \ldots\}$ and infinitesimals such as $1/\omega = \{0 | 1,\ \frac{1}{2},\ \frac{1}{4},\ \ldots\}$.

The value of a game may be thought of as "the number of moves advantage to Left". For example, -2 is two moves advantage to Right. The first four Blue–Red Hackenbush values in Figure 12 are clear. Deletion of the blue edge in the fifth reduces the picture to 0, while deletion of the red edge leaves 1, so the fifth value is $\{0 | 1\} = \frac{1}{2}$.

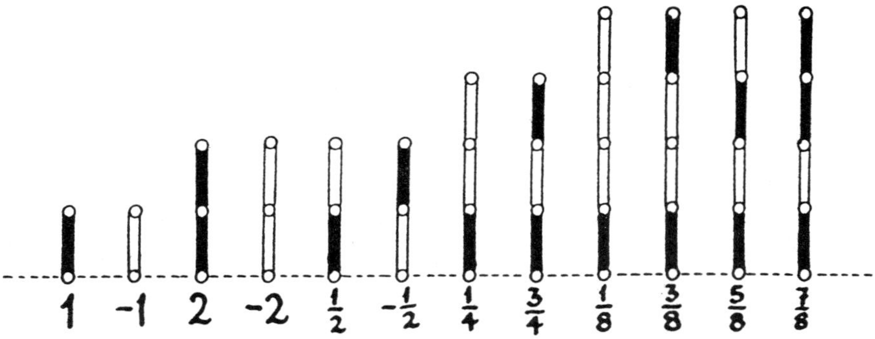

Figure 12: Values of some Blue–Red Hackenbush strings.

Check that if you play a game comprising two copies of this and a single separate red edge, Figure 13, then the second player wins (although if Left starts, Right *can* make a bad reply). Check the remaining values in Figure 12.

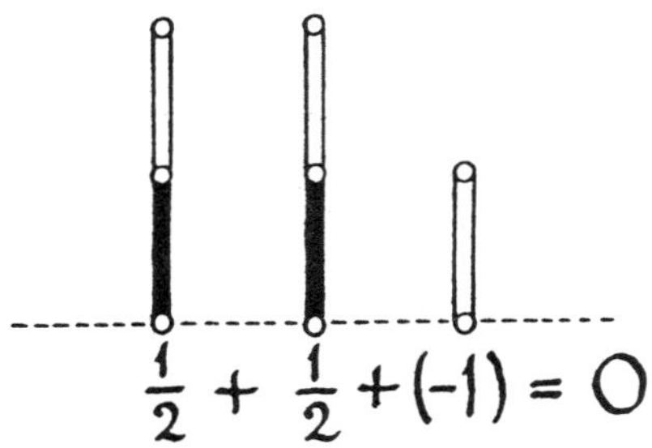

Figure 13: Two halves make a whole.

Not only is the value of every Blue–Red Hackenbush picture a number, but every number can be represented by a Hackenbush string! To see this, work backwards from **Berlekamp's Rule** for evaluating Hackenbush strings [Berlekamp (1974)]:

> The **sign** is determined by the color of the edge touching the ground (+ for blue, − for red). Move up the string until there is a change in edge color. This first *pair* of differently colored consecutive edges represents the **binary point**. The number of edges *below* this pair gives the **integer part** of the number. Above the pair, label each edge with a binary digit, 1 or 0, according as its color agrees or disagrees with the color of the edge touching the ground, and adjoin an extra digit 1 at the end if the string is finite.

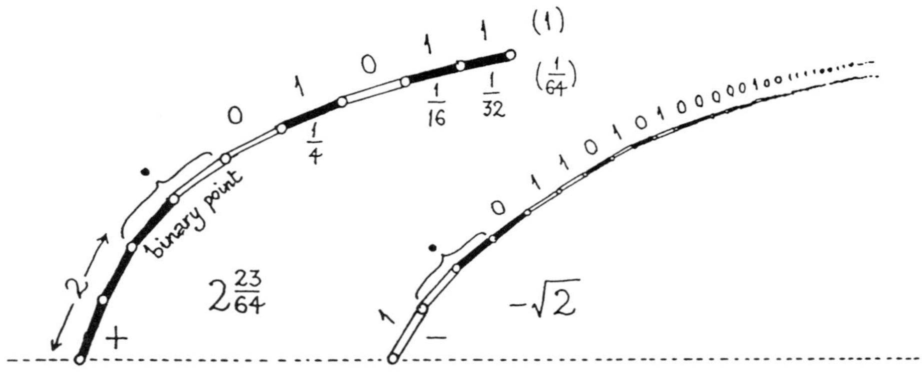

Figure 14: Berlekamp's Rule for Hackenbush strings.

The rule is illustrated in Figure 14; also use it to check the values in Figure 12. Infinitesimals and infinite ordinals can also be represented by (infinite) Hackenbush strings [WW, 309–313].

13 Simplifying Games

We may be able to do this in either of two ways [ONAG, 109–112; WW, 62–64]:

by deleting dominated options, or

by replacing reversible options.

We used the former implicitly when we made our catalog of day-two games. If in a game $G = \{A,\ B,\ C,\ \ldots\ |Z,\ Y,\ X,\ \ldots\}$ we have $B \geq A$ (resp. $Y \leq Z$), then B **dominates** A (resp. Y dominates Z) and A (resp. Z) may be deleted, provided that B (resp. Y) is retained.

Replacing reversible options is more subtle. A right option X is **reversible** if Left has a reply X^L which is at least as good for her as the original game, that is if $X^L \geq G$. Then X may be replaced by the list of *all* the Right options, $X^{LR} = \{X^{LR_1},\ X^{LR_2},\ \ldots\ \}$, of X^L. Similarly, the Left option C is reversible if Right has a reply C^R at least as good for him as G is, that is if $C^R \leq G$, and then C may be replaced by the list of all the Left options C^{RL} of C^R.

For example, in the game

$$G = \{0, *|*\},$$

which is a more precise description of the shape labelled "$\uparrow$" in Figure 10, neither Left option dominates the other, since $*\|0$, but the Left option $*$ is reversible, because $*^R = 0 \le G$. [To see that $G \ge 0$, note that if Right starts, his only option is $* = \{0|0\}$ and Left plays to 0 and wins.] So the Left option $*$ may be replaced by all the Left options of $*^R = 0 = \{\ |\ \}$; that is, it may be replaced by all the members of the empty set, that is, it may be deleted, and G simplifies to $\{0|*\} = \uparrow$.

For another example, consider the game $H = \{\uparrow\ |\ \uparrow\}$ and examine each option for reversibility (Figure 15).

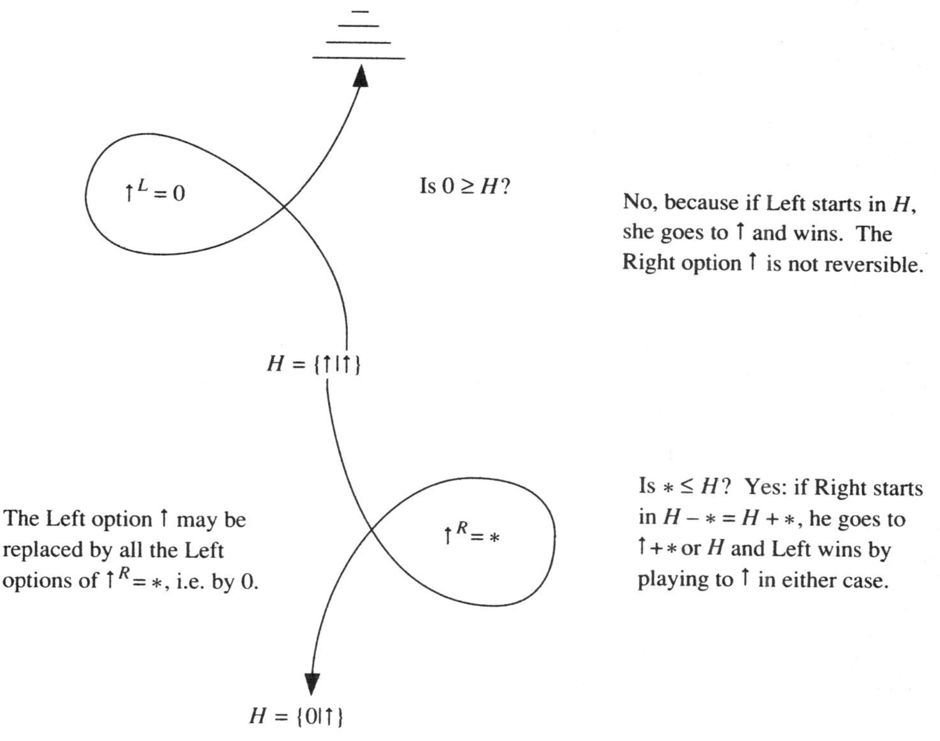

Figure 15: Examining $\{\uparrow\ |\ \uparrow\}$ for reversibility.

Check that H satisfies the **upstart equality** [ONAG, 77; WW, 73],

$$\{0|\uparrow\} = \uparrow + \uparrow + * = \Uparrow * (\text{"double up star"}).$$

Proceedings of Symposia in Applied Mathematics
Volume **43**, 1991

Numbers and Games

John Horton Conway

1 Introduction

Much of this chapter is taken from ONAG, chapters 0 and 1.

Let us see how those who were good at constructing numbers have approached this problem in the past.

Dedekind (and before him the author—thought to be Eudoxus—of the fifth book of Euclid) constructed the real numbers from the rationals. His method was to divide the rationals into two sets L and R in such a way that no number of L was greater than any number of R, and use this "section" to define a new number $\{L|R\}$ in the case that neither L had a greatest member nor R had a least.

His method has been criticized on several counts. Perhaps the most important is that the rationals are supposed to have been already constructed in some other way, and yet are "reconstructed" as certain real numbers. The distinction between the "old" and "new" rationals seems artificial but essential.

Cantor constructed the infinite ordinal numbers. Supposing the integers 1, 2, 3, ... given, he observed that their *order-type* ω was a new (and infinite) number greater than all of them. Then the order-type of $\{1, 2, 3, \ldots, \omega\}$ is a still greater number $\omega + 1$, and so on, and on, and on. The similar objections to Cantor's procedure have been met by von Neumann, who observes that it is unnecessary to suppose that 1, 2, 3, ... are already given, and that it is natural to start at 0 rather than 1. He also takes each ordinal as the *set* (rather than the order-type) of all previous ones. Thus for von Neumann, 0 is the empty set, 1 the set $\{0\}$,

The paper is in final form and no version of it will be submitted for publication elsewhere.
1980 *Mathematics Subject Classification* (1985 *Revision*). Primary 90D05.

2 the set {0,1}, ..., ω the set {0, 1, 2, ...}, and so on.

These two methods are part of a simpler, yet far more general method by which we can construct a very large Class **No** of numbers, including at the same time the real numbers and the ordinal numbers, and many others. Because of the generality of this Class, we shall simply describe its members as *numbers*, without adding any restricting adjective. Since our first discovery of the construction, Donald Knuth has suggested the happy name of **Surreal Numbers** [Addison-Wesley, 1974]. It will turn out that **No** is a Field (i.e. a field whose domain is a proper Class)—in general we shall capitalize the initial letter of any "big" concept, on the grounds that proper Classes, like proper names, deserve capital letters. So, for instance, the word *Group* will mean any group whose domain is a proper class.

2 All Numbers Great and Small

If L, R are any two sets of numbers, and no member of L is $\geq$ any member of R, then there is a number $\{L|R\}$. All numbers are constructed in this way.

Notice that when the condition on L and R is omitted, we obtain the more general notion of a **game**. For example, in the book *Surreal Numbers* Alice discovers what she calls a **pseudonumber**, namely $\{1|0\}$, and she was able to show that it was $\leq$ a number y if and only if $y > 1$ and that it is $\geq$ a number x if and only if $x < 0$. "It's not related at all to any numbers between 0 and 1." This is an example of a "hot game", which has a "confusion interval" of length one.

Revenons à nos moutons!

If $x = \{L|R\}$ we write x^L for the typical member of L, and x^R for the typical member of R. For x itself we then write $\{x^L|x^R\}$. Although x^L and x^R are numbers, it's advisable to think of them as (typical) *members of sets.*

$x = \{a, b, c, \ldots |d, e, f, \ldots\}$ means that $x = \{L|R\}$, where $a, b, c, \ldots$ are the members of L, and $d, e, f, \ldots$ are the members of R.

Definition of $x \geq y$, $x \leq y$.

We say $x \geq y$ just if (no $x^R \leq y$ and $x \leq$ no y^L), and $x \leq y$ just if $y \geq x$.
We write $x \not\leq y$ to mean that $x \leq y$ does not hold.

Definition of $x = y$, $x > y$, $x < y$.

$x = y$ just if ($x \geq y$ and $y \geq x$).

$x > y$ just if ($x \geq y$ and $y \ngeq x$).

$x < y$ just if $y > x$.

Definition of $x + y$

$$x + y = \{x^L + y, x + y^L | x^R + y, x + y^R\}.$$

Definition of $-x$

$$-x = \{-x^R | - x^L\}.$$

Definition of xy

$$xy = \{x^L y + xy^L - x^L y^L, x^R y + xy^R - x^R y^R | x^L y + xy^R - x^L y^R, x^R y + xy^L - x^R y^L\}.$$

It is remarkable that these eight statements already define a real-closed Field with a very rich structure. Notice that *equality is a defined relation* so that different-looking definitions can produce the same number! We must distinguish between the *form* $\{L|R\}$ of a number and the number itself.

All the definitions are inductive, so that to decide, for example, whether $x \geq y$, we must consider a number of similar questions about the pairs x^R, y and x, y^L. But each of these problems is simpler than the given one. It may not be quite so obvious that the inductions require no basis, but ultimately we are reduced to problems about members of the empty set.

What are the motives behind our definitions? We want each new number x to lie between the numbers x^L (to the left) and x^R (to the right), and that $\geq$, $+$, $-$, $\cdot$, etc., shall have their usual properties. If, for example, $y \geq$ some x^R we would not have $x \geq y$, for then $x \geq x^R$. Similarly we could not allow $x \geq y$ if $x \leq$ some y^L. So we define $x \geq y$ in all other cases. This conforms with our motto

Whatever is not forbidden, is permitted. *J.C.F. von Schiller*

and helps to ensure that numbers are totally ordered.

The spirit of the definitions is to ask what we know already (by the answers to *simpler* questions) about the object being defined, and to make the answers part of the definition. Thus if addition is to have nice properties and if x is between x^L and x^R, and y is between y^L and y^R, then we know "already" that $x + y$ must lie between both $x^L + y$ and $x + y^L$ (on the left) and $x^R + y$ and $x + y^R$ (on the right), which yields the definition of $x + y$. Similarly $-x$ will lie between $-x^R$ (on the left) and $-x^L$ (on the right), which suffice to define $-x$. It was much harder to find the definition of xy, but we can summarize our comments by saying that the definitions of the various operations and relations are just the simplest possible definitions which are consistent with their intended properties. The definition of the product assures that it is the simplest function which satisfies what Martin Kruskal has called the **rectangle inequality**

$$a > b, \quad c > d \quad \Rightarrow \quad ac + bd > ad + bc$$

(the area of a rectangle is positive).

3 Examples of Numbers

Our construction gets off the ground by taking L and R in $\{L|R\}$ each to be *the empty set*, $\varnothing$. We call the result **zero** and write 0. Is 0 a number? Is $0 \geq 0$? Yes, yes, and happily $0 = 0$. Verify that $-0 = 0 + 0 = 0$.

We can now use the sets $\varnothing$ and $\{0\}$, and we get $\{\varnothing \,|\, \varnothing\}$, $\{0|\varnothing\}$, $\{\varnothing\,|0\}$ and $\{0|0\}$. The first is zero, of course, and the last is not a number: why not? But the other two are numbers (check!) which we call

$$1 = \{0| \quad \} \qquad \text{and} \qquad -1 = \{ \quad |0\}$$

Verify that $0 \not\geq 1$, $1 \geq 0$, $1 > 0$, $-1 < 0$, $-1 \not\geq 1$, and that $1 \geq 1$.

We now have three numbers $-1 < 0 < 1$ and so $2^3 = 8$ possible sets for L and R. But the condition that no member of L should be $\geq$ any member of R restricts us to eight possibilities $\{ \quad |R\}$, eight possibilities $\{L| \quad \}$ and the five

$$\{-1|0\},\ \{-1|0,1\},\ \{-1|1\},\ \{0|1\}, \quad \text{and} \quad \{-1,0|1\}.$$

Verify that of these twenty-one, four are repeats; $\{ \quad |1\}$, $\{-1| \quad \}$ and $\{-1|1\}$ are new disguises for zero; $\{-1,0| \quad \}$ and $\{ \quad |0,1\}$ new forms of 1 and -1, and the rest define four new numbers:

$$\{1| \quad \} = \{-1,1| \quad \} = \{0,1| \quad \} = \{-1,0,1| \quad \}$$
$$\{ \quad |-1\} = \{ \quad |-1,0\} = \{ \quad |-1,1\} = \{ \quad |-1,0,1\}$$
$$\{0|1\} = \{-1,0|1\} \quad \text{and} \quad \{-1|0\} = \{-1|0,1\}$$

Check these equalities and check that if we give these the names 2, -2, $\frac{1}{2}$ and $-\frac{1}{2}$, then these numbers have the properties that we have come to expect of them. Prove that $x + 0 = x = 0 + x$ and that $1 + 1 = 2$. Can you prove that $\frac{1}{2} + \frac{1}{2} = 1$? This gets quite tedious: look on page 9 of ONAG if you haven't the stamina to push it through for yourself.

If we have to keep going back to first principles, things get increasingly tedious, but we begin to suspect that the numbers are totally ordered, and that we need only consider the greatest x^L (if any) and the least x^R (if any). Then, on the next "day" (the third) there will only be four new numbers:

$$0 < \{0|\frac{1}{2}\} < \frac{1}{2} < \{\frac{1}{2}|1\} < 1 < \{1|2\} < 2 < \{2|\ \}$$

and their negatives. To cut a long story short, these are the numbers

$$\frac{1}{4}, \quad \frac{3}{4}, \quad 1\frac{1}{2} \quad \text{and} \quad 3.$$

The general pattern of "births" is exhibited in Figure 1 (taken from WW, p. 25) where the numbers are related to the values of Blue-Red Hackenbush strings. Figure 2 is from page 11 of ONAG, and shows a few of the numbers born on days ω, $\omega+1$, $\ldots$, $\omega.2$. Figure 3 is the frontispiece of ONAG, and gives a hint of just how much further you can go.

On day ω the numbers $\pm\omega$ and *all* the real numbers are born. Also some infinitesimals, such as $1/\omega = \{0|1, \frac{1}{2}, \frac{1}{4}, \frac{1}{8}, \ldots\}$. The next day we get $\{0, 1, 2, \ldots, \omega|\ \} = \omega + 1$, of course, but $\{0, 1, 2, \ldots|\omega\} = \omega - 1$ as well. Then $\{0, 1, 2, \ldots|\omega, \omega - 1\} = \omega - 2$ and so on to

$$\{0, 1, 2, \ldots|\omega, \omega - 1, \ldots, \omega - (n-1)\} = \omega - n,$$

$$\{0, 1, 2, \ldots|\omega, \omega - 1, \omega - 2, \ldots\} = \omega/2.$$

and you may be able to construct $\omega/3$ and believe that

$$\{0, 1, 2, \ldots|\omega, \omega/2, \omega/4, \omega/8, \ldots\}$$

is a square root of ω.

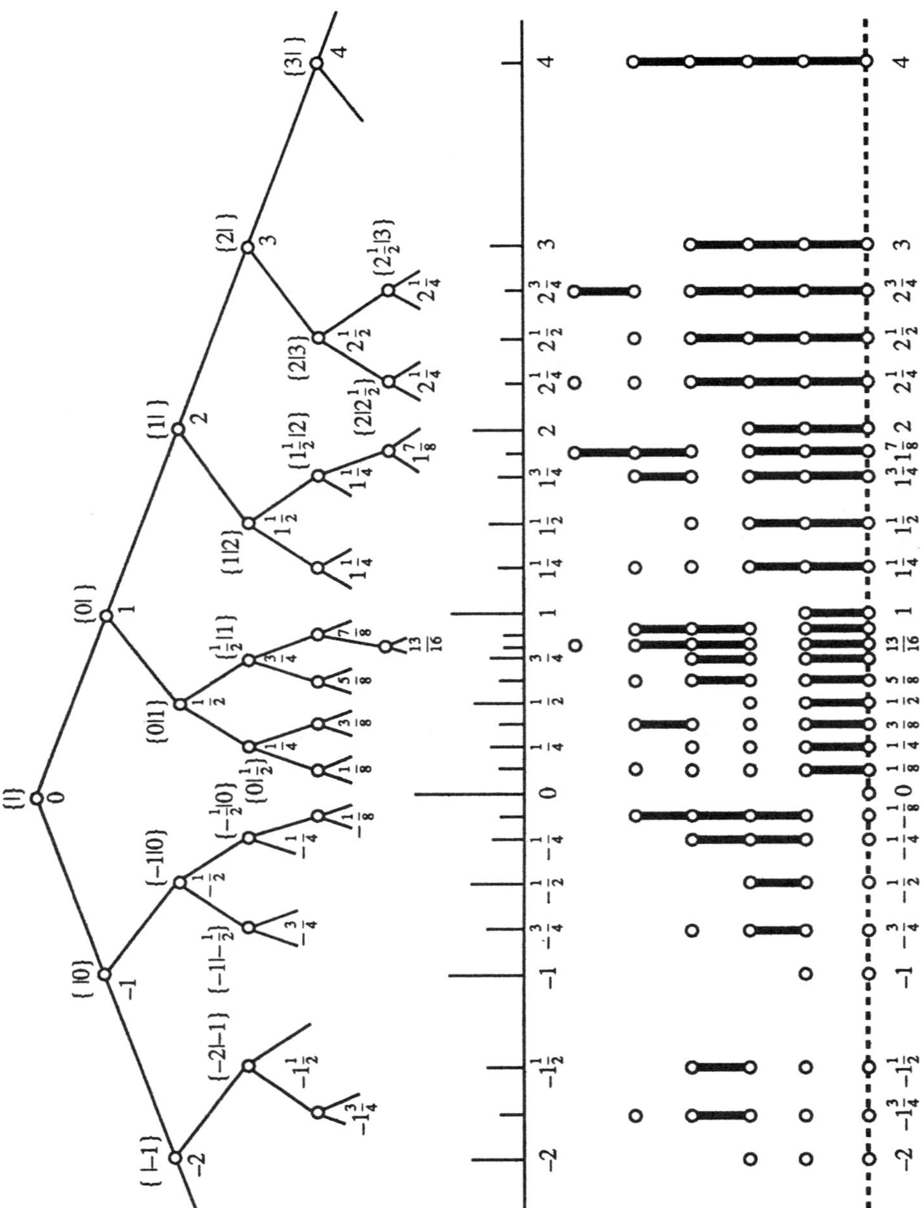

Figure 1: Australian number tree, the real number line, and Hackenbush strings.

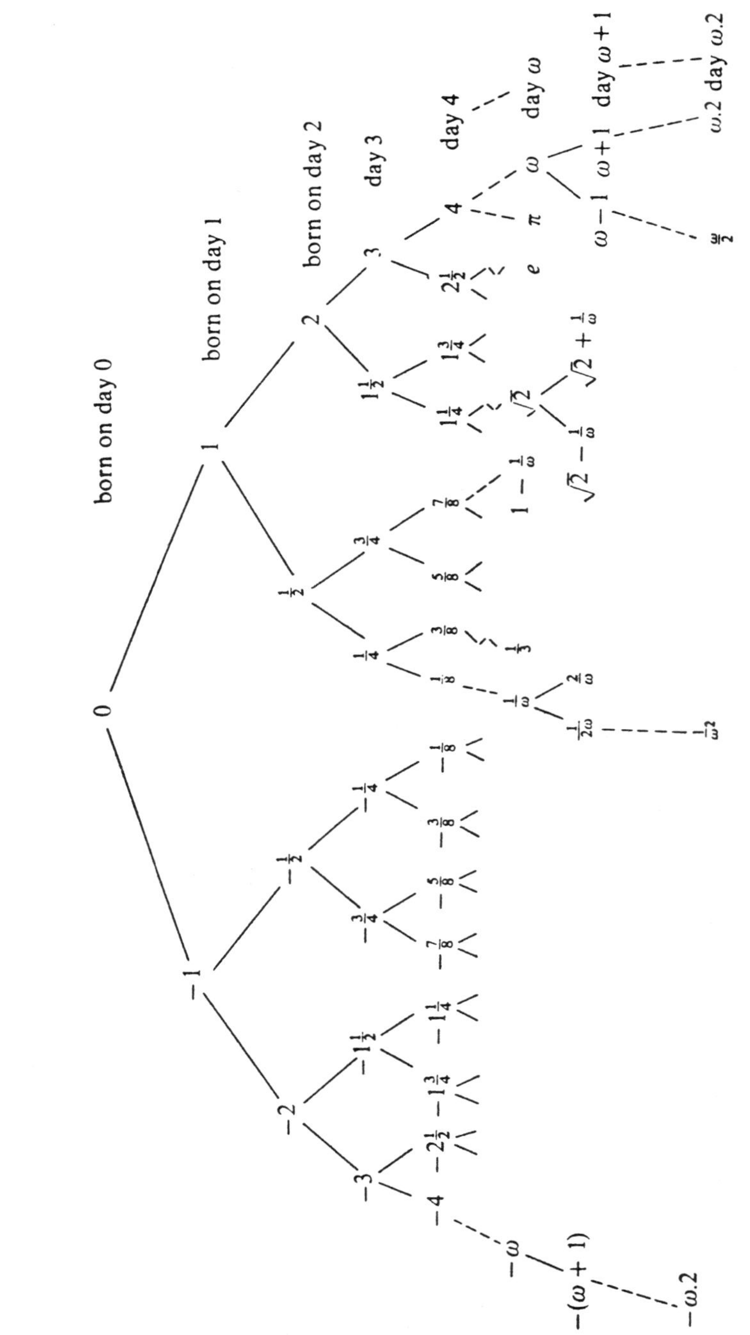

Figure 2: When the first few numbers were born.

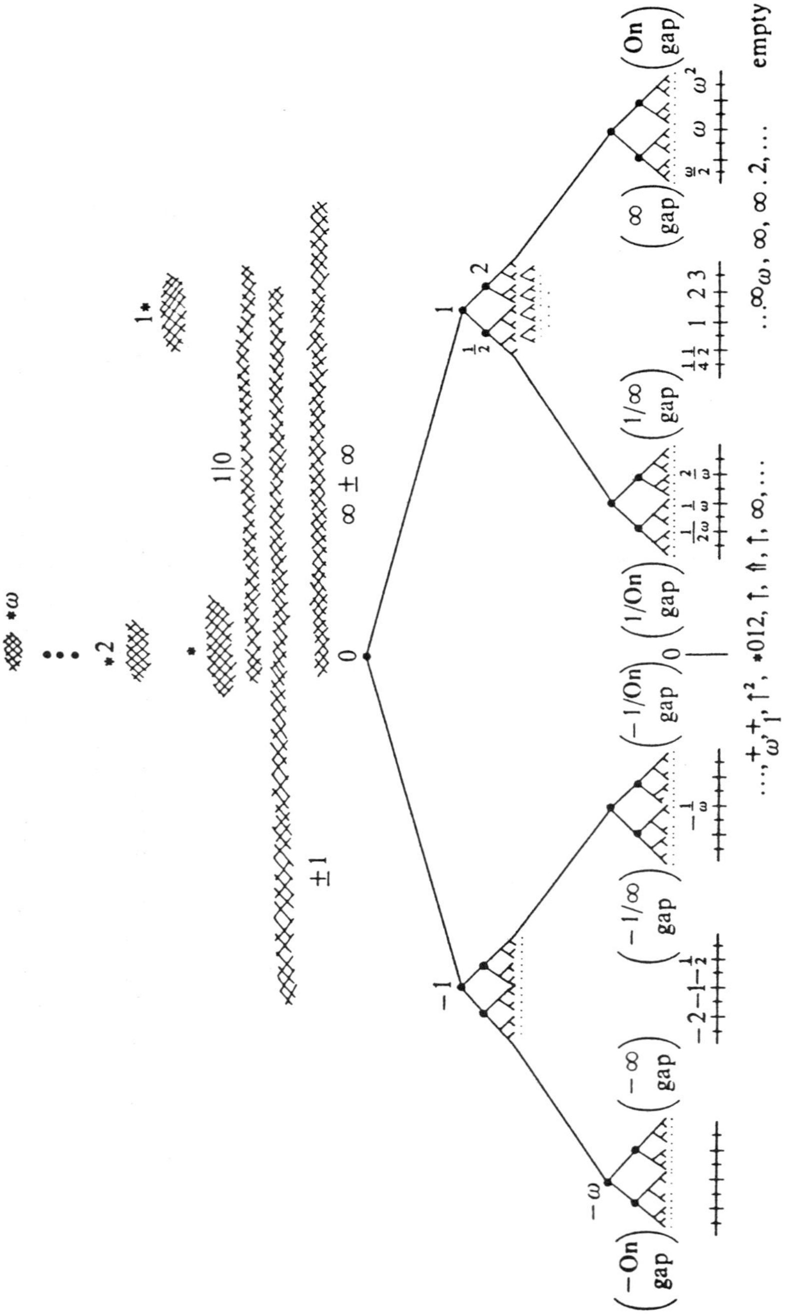

Figure 3: The tree of numbers, and the positions of some games.

4 Games

The construction for numbers generalizes immediately to a construction for **games**.

If L and R are any two sets of games, then there is a game $\{L|R\}$. All games are constructed in this way.

The order-relations and arithmetic operations on games are defined by the same definitions as for numbers. The most important distinction between numbers and more general games is that numbers are totally ordered, but games are not—there are games x and y for which we have *neither* of $x \geq y$, $y \geq x$.

To show that a game $x = \{x^L|x^R\}$ is a number, we must show *firstly* that all of the games x^L, x^R are numbers, and *secondly* that there is no inequality of the form $x^L \geq x^R$.

5 Identity and Equality

We call games x and y **identical** ($x \equiv y$) if their left and right sets are identical—every x^L identical to some y^L, every x^R identical to some y^R and vice versa. Recall that x and y are defined to be **equal** ($x = y$) just if we have both $x \geq y$ and $y \geq x$. The distinction causes no great problems until we come to multiplication, where the trouble is that there can exist *equal* (but not, of course, identical) games x and y for which xz and yz are unequal!

6 Order and Equality

Theorem 0 *For all games x we have (i) $x \not\geq x^R$, (ii) $x^L \not\geq x$, (iii) $x \geq x$, (iv) $x = x$.*

Theorem 1 *If $x \geq y$ and $y \geq z$, then $x \geq z$.*

Theorem 2 *For any number x we have $x^L < x < x^R$ for all x^L, x^R. Also, for any two numbers x and y we must have $x \leq y$ or $x \geq y$.*

7 Addition, Negation and Order

Theorem 3 *For all x, y, z we have*

$$x + 0 \equiv x, \qquad x + y \equiv y + x, \qquad (x + y) + z \equiv x + (y + z).$$

Theorem 4 *(i)* $-(x + y) \equiv -x + -y$, *(ii)* $-(-x) \equiv x$, *(iii)* $x + -x = 0$.

Note that (iii) is an equality, rather than an identity. With equality rather than identity, addition is a commutative Group operation, with 0 for zero and $-x$ for the negative of x. This is true for general games.

Theorem 5 *We have $y \geq z$ just if $x + y \geq x + z$.*

Theorem 6 *(i) 0 is a number, (ii) if x is a number, so is $-x$, (iii) if x and y are numbers, so is $x + y$.*

If we continued with the properties of multiplication and division, we could show that

The Class **No** of all numbers forms a totally ordered Field.

8 The game-theoretic proofs

What we have described so far is a very large Class of numbers obtained by generalizing the idea of Dedekind cut, and in which the arithmetical notions are defined by rather abstract transfinite inductions. When we widen the theory to games these arithmetical notions (except multiplication) have strategic interpretations which make the proofs trivial.

For, the notation $G = \{G^L | G^R\}$ now indicates a game from which Left's typical move is to G^L, Right's to G^R, and so

$$G + H = \{G^L + H, G + H^L | G^R + H, G + H^R\}$$

defines the notion of **disjunctive sum**, wherein a player chooses either component at will and makes a legal move in it, leaving the other component untouched. Similarly

$$-G = \{-G^R | -G^L\}$$

indicates the **opposite game**, obtained by interchanging the roles of Left and Right throughout. We say that $G \geq H$ provided that, for all X, and for all choices of who is to start,

if Left can win $H + X$, then Left can win $G + X$.

It is now easy to check that we have

$$\begin{array}{ccccc} G > 0 & G < 0 & G = 0 & G \parallel 0 & \text{according as} \\ L & R & 2 & 1 & \end{array}$$

can always win G (here '1' and '2' denote the first and second player to move), and the most important property of all:

we have $G \geq H$ just if Left can win $G - H$ when Right starts.

The easy proofs are all by combining strategies. For example the assertion

$$G \geq 0, \quad H \geq 0 \quad \Rightarrow \quad G + H \geq 0$$

means that Left can win $G + H$ when Right starts provided only that he can win each of G and H when Right starts, and he does this by responding to any move in G or H by the response given by his strategy for that game.

9 What numbers tell us about games

The first thing is that numbers form a subClass of games closed under the arithmetic operations. Moreover this subClass is totally ordered and its arithmetic is well understood. The simplest few positions in *any* game are likely to be numbers, and they help us to get the analysis "off the ground", so to speak.

There are some more subtle relations between numbers and general games, however, and most of these are conveniently conveyed by the **thermograph** of a game G, a diagram that shows us how the cooled forms G_t of G compare with all numbers. The **cooled game** G_t is defined by

$$G_t = \{G_t^L - t | G_t^R + t\}$$

except that if t is greater than the smallest t ($= t_0$, say) for which the right hand side is infinitely close to some number (m, say) then $G_t = m$ for all larger t.

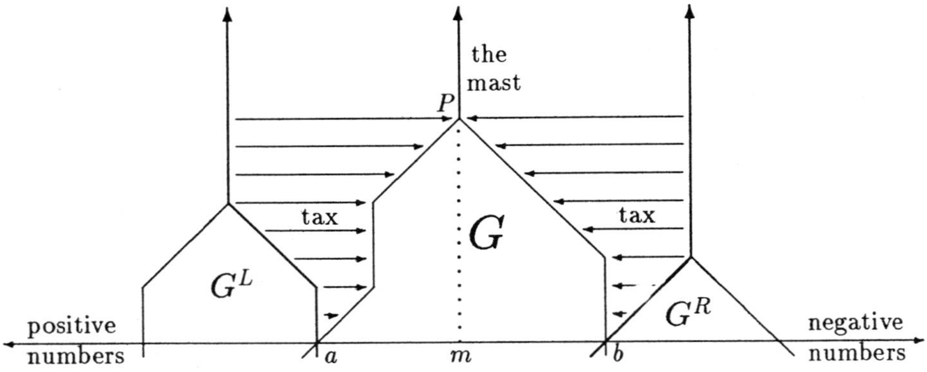

Figure 4: The thermograph of $G = \{G^L | G^R\}$.

The figure shows how to compute the thermograph for a case where G has just one Left option and just one Right option. We take the right boundary of G^L and "tilt" it right by an amount t at height t and also we take the left boundary of G^R and tilt it left in a similar way. The two resulting "curves" define the left and right boundaries of the thermograph of G until they meet (at P, at height t_0 in the figure) above which they are replaced by a vertical mast.

If G had more than one Left option, we'd take the *leftmost right boundary* at each height. For more precise details, see ONAG, 104–108 or WW, 149–181.

The thermograph shows how G compares with all numbers. For the example of Fig. 4, G is confused with all numbers $> b$ and $\leq a$ and comparable with all other numbers.

Also m is called the **mean value** of G and t_0 its **temperature**. We have

$$nm - t_0 - \epsilon < nG < nm + t_0 + \epsilon$$

for all integers n and all numbers $\epsilon > 0$.

Proceedings of Symposia in Applied Mathematics
Volume **43**, 1991

Impartial Games

Richard K. Guy

1 Introduction

Alternative treatments of impartial games are given in ONAG, 112–130 and in WW, 81–116; and an elementary introduction in [**3**].

In the first chapter we defined an **impartial** game as one in which the set of Left options is the same as the set of Right options. We've already noticed the impartial games

$$\{\ \mid\ \} = *0 = 0, \qquad \{0|0\} = *1 = * \quad \text{and} \quad \{0, *|0, *\} = *2.$$

that were born on days 0, 1, and 2 respectively, so it should come as no surprise that on day n, the game

$$*n = \{*0, *1, *2, \ldots, *(n-1) | *0, *1, *2, \ldots, *(n-1)\}$$

is born. In fact any game of the type

$$\{*a, *b, *c, \ldots | *a, *b, *c, \ldots\}$$

has value $*m$, where $m = \operatorname{mex}\{a, b, c, \ldots\}$, the least non-negative integer *not* in the set $\{a, b, c, \ldots\}$. To see this, notice that any option $*n$ with $n > m$ is reversible, because the option $*m$ of $*n$ is $\geq *m$ and so $*n$ may be deleted in favor of the options of $*m$, namely 0, $*$, $*2$, ..., $*(m-1)$.

This is the inductive step which proves the Sprague-Grundy theorem [Sprague (1935-36), Grundy (1939)], which states that every (position in an) impartial game is equivalent to a **nim-heap**.

As the Left and Right options are the same, $*n$ is its own negative, $*n + *n = 0$. Also, we need only write one set of options, and may define

The paper is in final form and no version of it will be submitted for publication elsewhere.
1980 *Mathematics Subject Classification* (1985 *Revision*). Primary 90D05.

the **nimber**

$$*n = \{0, *, *2, \ldots, *(n-1)\}.$$

This exactly parallels John von Neumann's [1923] definition of ordinal numbers.

2 Examples of Impartial Games

As was mentioned in the first chapter, the game of **Nim** is played with several heaps of beans. A move is to select a heap, and to remove any positive number of beans from it, possibly the whole heap. Any position in Nim is therefore the sum of several one-heap Nim games. The value of a single heap of n beans is $*n$.

It's easy to see how to win a game of Nim if there's only one (non-empty) heap: take the whole heap! But it's worth noting exactly what your options are. They are to move to any smaller sized heap: they correspond exactly to the options in the definition of $*n$. It's also fairly easy to play well at two-heap Nim: if the heaps are unequal in size, remove enough beans from the larger to make the heaps equal. If the two heaps are already equal, then hope that it is your opponent's turn to move. From then on, use the **Tweedledum and Tweedledee Principle**: whatever your opponent does to one heap, you copy in the other. For more than two heaps, the theory is more tricky. It was discovered by Bouton (1902). Imagine each heap to be partitioned into distinct powers of two. For example, Figure 1 shows four heaps of 27, 23, 22 and 15 beans partitioned in this way.

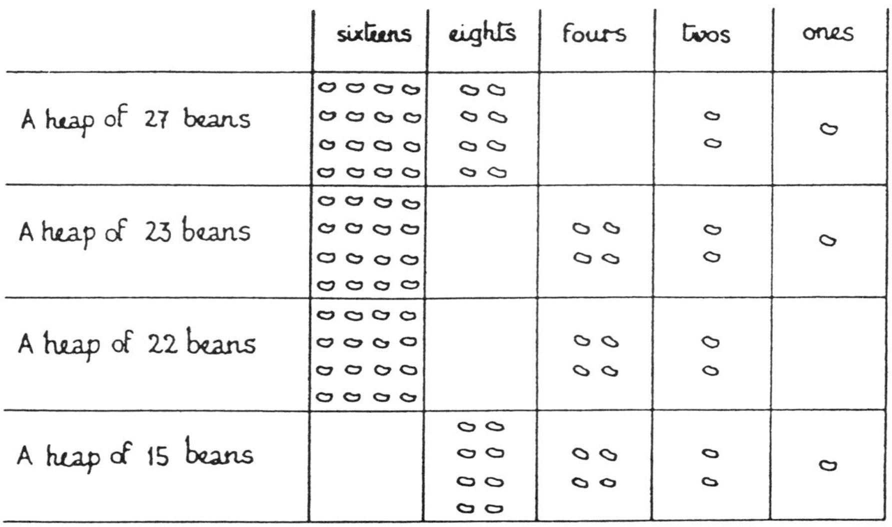

Figure 1: How to look at a Nim position.

We can ignore heaps of equal sizes, so concentrate on the columns with an *odd* number of parts, the ones, fours and sixteens. A good move would be to take $16 + 4 + 1 = 21$ from the 23 heap. If you leave a position for your opponent in which each power of two occurs *evenly* often, then he will have to change the parity in at least one column, and then you will be able to restore it. Notice that finding a good move does *not* depend on their being appropriate powers of two all in the same row (heap). You could also take $16 - 4 + 1 = 13$ from the 27 heap, or $16 + 4 - 1 = 19$ from the 22 heap. Find three good moves from the Nim position {23, 19, 13, 12, 11}.

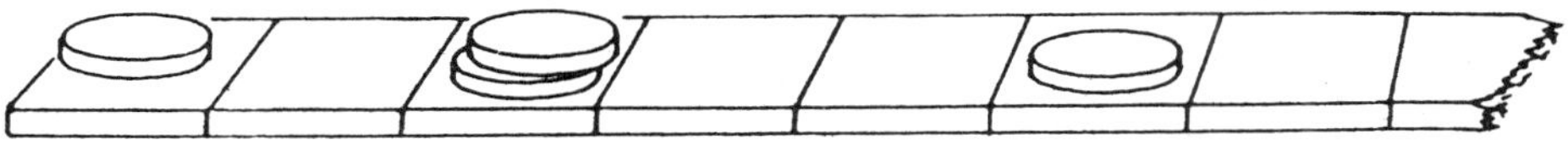

Figure 2: A game of Nimble.

Nimble is played with coins on a strip of squares (Figure 2). Take turns, moving just one coin to the left. You can jump onto or over other coins, even clear off the strip. You can have any number of coins on a square. The last player wins. Can you analyze this game? Suppose that you are not allowed to jump off the strip, so that the game ends when all coins are stacked on the left hand square. Can your analysis be modified to cope with this variant?

Figure 3: A coin-sliding game.

In the game shown in Figure 3, you're allowed at most one coin on a square, and you're not allowed to jump over other coins. A move is to *slide* a coin leftwards as far as you like, but not onto or over the next coin, and not off the end of the strip. The analysis is now more cunning: the black marks on the side of the strip may give you a hint.

Figure 4 shows N.G. de Bruijn's **Silver Dollar Game**, which is played

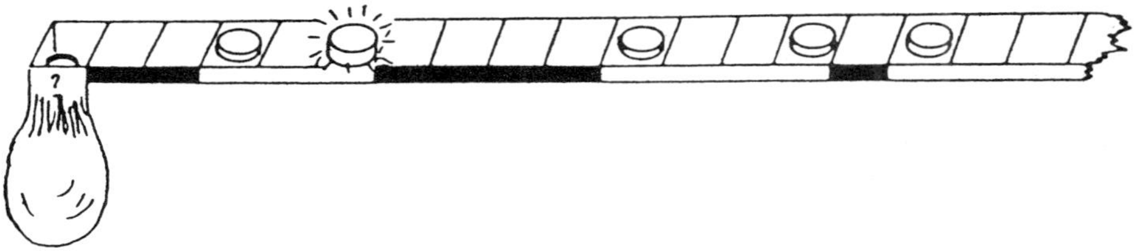

Figure 4: De Bruijn's Silver Dollar Game.

like the previous game, but one coin is worth much more than all the others put together, the leftmost square is replaced by a money-bag, and there's the additional option of taking the money-bag. If you do this, your opponent gets the coins left on the strip. When you've solved that game, consider the variant in which the additional option is to slide a coin *and* take the money-bag, all in one move.

Poker Nim is played like Nim, but with poker chips in place of beans; as well as removing chips from a heap, you may instead *add* chips to a heap. How does this affect play?

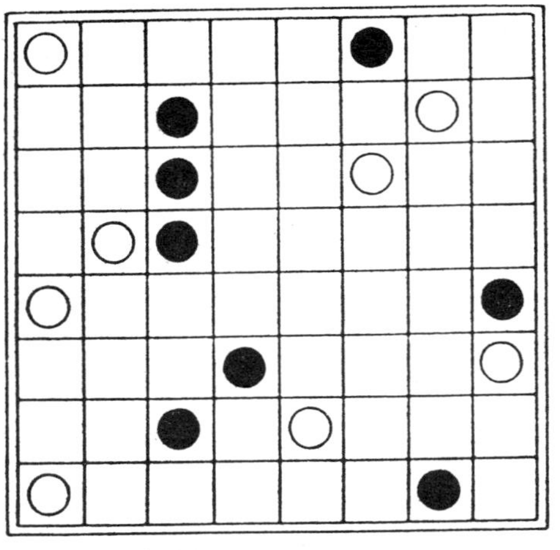

Figure 5: Northcott's Game.

In **Northcott's Game** there is one checker of each color on each row of a checker-board (Figure 5). A move is to slide one of your checkers any number of squares in its own row, without jumping over your opponent's checker and without going off the board. This *looks* like a *partizan* game, and many people can't see any point to it. Remember that the aim is to be the last player to move.

Lasker's Nim is played like ordinary Nim, but with the additional option that you are allowed to split a heap into two smaller, non-empty heaps.

3 Nim-Addition

[ONAG, 50–51; WW, 60–61]

We know that $*n + 0 = *n$ and that $*n + *n = 0$, and it's not hard to see that addition of impartial games, indeed of any of our games, is commutative and associative. Let's calculate

$$*2 + * = \{0, *\} + \{0\} = \{0 + *, * + *, *2 + 0\} = \{*, 0, *2\} = *3.$$

Add $*$, or $*2$, to each side and obtain $*2 = *3 + *$ and $* = *3 + *2$. In general,

$$\begin{aligned} *a + *b &= \{0, *, *2, \ldots, *(a-1)\} + \{0, *, *2, \ldots, *(b-1)\} \\ &= \{0+*b, *+*b, \ldots, *(a-1)+*b, *a + 0, *a + *, \ldots, *a + *(b-1)\} \end{aligned}$$

and we can build a nim-addition table (Table 1) by noting that the options of an entry are just the earlier entries in the same row and the earlier entries in the same column. Each entry in Table 1 is the least non-negative integer not appearing as an earlier entry in the same row or column. For instance, $*5 + *6 = *3$, because 3 is the first number not in the set $\{5, 4, 7, 6, 1, 0, 6, 7, 4, 5, 2\}$, i.e., the first six entries in row 5 and the first five entries in column 6. In the usual language, 3 is the **nim-sum** of 5 and 6, which is sometimes written $5 \overset{*}{+} 6 = 3$.

0	1	2	3	4	5	6	7	8	9	10	11	12	13	14	15
1	0	3	2	5	4	7	6	9	8	11	10	13	12	15	14
2	3	0	1	6	7	4	5	10	11	8	9	14	15	12	13
3	2	1	0	7	6	5	4	11	10	9	8	15	14	13	12
4	5	6	7	0	1	2	3	12	13	14	15	8	9	10	11
5	4	7	6	1	0	3	2	13	12	15	14	9	8	11	10
6	7	4	5	2	3	0	1	14	15	12	13	10	11	8	9
7	6	5	4	3	2	1	0	15	14	13	12	11	10	9	8
8	9	10	11	12	13	14	15	0	1	2	3	4	5	6	7
9	8	11	10	13	12	15	14	1	0	3	2	5	4	7	6
10	11	8	9	14	15	12	13	2	3	0	1	6	7	4	5
11	10	9	8	15	14	13	12	3	2	1	0	7	6	5	4
12	13	14	15	8	9	10	11	4	5	6	7	0	1	2	3
13	12	15	14	9	8	11	10	5	4	7	6	1	0	3	2
14	15	12	13	10	11	8	9	6	7	4	5	2	3	0	1
15	14	13	12	11	10	9	8	7	6	5	4	3	2	1	0

Table 1. Nim-addition table. The stars have been omitted; i.e., the entries are nim-values instead of nimbers.

Contrast the two equations $*5 + *6 = *3$ and $5 \overset{*}{+} 6 = 3$. In the first the summands are nimbers, i.e., values of impartial games, and the addition is a game sum. In the second the summands are **nim-values** and the addition is nim-addition.

Nim-addition is perhaps better known as addition without carry in base 2, or vector or coordinatewise addition over $GF(2)$, or XOR (exclusive or): it is reassuring that it also follows from the more general idea of game sum.

Many of the games mentioned in the previous section are disguises for Nim, often with the addition of **reversible moves**, which we disposed of in the first lecture. As we saw in the Introduction, every impartial game is equivalent to a **bogus nim-heap**, i.e., a heap of m beans ("m" for "mex"), together with some (reversible) options which *increase* the size of the heap.

To summarize the Sprague-Grundy theory: the nim-value of the sum of two impartial games is the nim-sum of their separate nim-values. Impartial games belong to one of only two outcome classes: all positions are either

$\mathcal{P}$-positions	previous-player-winning	nim-value zero, or
$\mathcal{N}$-positions	next-player-winning	nonzero nim-value.

In the literature, $\mathcal{P}$-positions are sometimes called "safe" or "good" or "winning" without indicating which player enjoys this happy situation.

4 Subtraction Games

[WW, 83–86, 487–498]

These are very simple examples of impartial games, played, like Nim, with heaps of beans. A move in the game $S(s_1, s_2, s_3, \ldots)$ is to take a number of beans from a heap, provided that number is a member of the **subtraction-set**, $\{s_1, s_2, s_3, \ldots\}$. Analysis of such a game and of many other heap games is conveniently recorded by a **nim-sequence**,

$$n_0 n_1 n_2 n_3 \ldots,$$

meaning that the nim-value of a heap of h beans is n_h, $h = 0, 1, 2, \ldots$, i.e., that the value of a heap of h beans in this particular game is the nimber $*n_h$. In this section, and often later, to avoid printing stars, we say that the nim-value of a position is n, meaning that its value is the nimber $*n$.

Table 2 shows some examples: the first is a manifestation of She-Loves-Me-She-Loves-Me-Not; the last is Nim. If the subtraction-set is finite, the nim-sequence is (ultimately) periodic. But little is known about the length of the period *vis à vis* the membership of the subtraction set.

Subtraction game	Nim-sequence	(ultimate) Period
$S(1)$	$\dot{0}\dot{1}01010101\ldots$	2
$S(2)$	$\dot{0}01\dot{1}0011001100\ldots$	4
$S(3)$	$\dot{0}0011\dot{1}0001110001110\ldots$	6
$S(1,2)$	$\dot{0}1\dot{2}0120120120\ldots$	3
$S(1,2,3)$	$\dot{0}12\dot{3}0123012301230\ldots$	4
$S(1,2,3,4)$	$\dot{0}123\dot{4}0123401234012340\ldots$	5
$S(2,4,7)$	$00112203\dot{1}0\dot{2}10210210\ldots$	3
$S(2,5,6)$	$\dot{0}011021302\dot{1}0011021302100\ldots$	11
$S(4,10,12)$	$\dot{0}00011110022113300221\dot{1}0000\ldots$	22
$S(1,2,3,4,\ldots)$	$0123456789\ldots$	(saltus 1 &) 1

Table 2. Nim-sequences and periods for subtraction games.

In subtraction games the nim-values 0 and 1 are remarkably related by **Ferguson's Pairing Property** [Ferguson (1974); WW, 86, 422]: if

s_1 is the least member of the subtraction-set, then

$$\mathcal{G}(n) = 1 \quad \text{just if} \quad \mathcal{G}(n - s_1) = 0.$$

Here and later "$\mathcal{G}(n) = v$" means that the nim-value of a heap of n beans is v.

5 Take-and-Break Games

[WW, 81–106]

Guy & Smith (1956) devised a code classifying a broad range of impartial games played with heaps or rows. If the binary expansion of the kth **code digit**, $\mathbf{d}_k$, in the game with code

$$\mathbf{d}_0{\cdot}\mathbf{d}_1\mathbf{d}_2\mathbf{d}_3\ldots$$

is

$$\mathbf{d}_k = 2^{a_k} + 2^{b_k} + 2^{c_k} + \ldots$$

where $0 \le a_k < b_k < c_k < \ldots$, then it is legal to remove k beans from a heap, provided that the rest of the heap is left in exactly a_k or b_k or c_k or ... non-empty heaps.

In order that the game should satisfy the ending condition, $\mathbf{d}_0$ must be divisible by 4, i.e., $a_0 \ge 2$.

Subtraction games are the special case $\mathbf{d}_s = \mathbf{3}$ when s is in the subtraction-set, and $\mathbf{d}_k = \mathbf{0}$ otherwise.

Octal games are those with code digits $\mathbf{d}_k \le \mathbf{7}$ for all k. Guy & Smith showed that an octal game is **ultimately periodic** with period p, i.e.,

$$\mathcal{G}(n+p) = \mathcal{G}(n) \quad \text{for all} \quad n > n_0 = 2e + p + t,$$

provided that $\mathcal{G}(n+p) = \mathcal{G}(n)$ for $n \le n_0$ apart from some exceptional values of n, of which e is the largest, and $\mathbf{d}_k = 0$ for $k > t$, i.e., the maximum number of beans that may be taken from a heap in a single move is t. Whether all such finite octal games are ultimately periodic remains a difficult open equation. They cannot be **arithmetically periodic**: that is, there is no period p and **saltus** $s > 0$, such that $\mathcal{G}(n+p) = \mathcal{G}(n) + s$ for all large enough n [WW, 114].

Table 3 exhibits a dozen specimen games, of which the last three are

Code	Game	Nim-sequence
·77	**Kayles.** Knock down one skittle, or two contiguous skittles, from a row. [Dudeney (1907), Loyd (1914)].	Ultimate period $p = 12$, 412814721827 except for $n = 0,3,6,9,11,15,18,21,22,28,$ 34,39,57,70, nim-value is resp. 0,3,3,4, 6,7,3,4,6,5,6,3,4,6.
·137	**Dawson's Chess.** $3 \times n$ board. White & Black pawns on ranks 1 & 3. Capturing obligatory. Looks partizan but isn't. [Dawson (1934, 1935)].	8112031103322445593301130211045374 except 0 for $n = 0, 14, 34$ & 2 for $n = 16, 17, 51$. $p = 34$.
·07	**Dawson's Kayles.** Knock down 2 skittles, but only if they're contiguous.	As for ·**137**, but shifted one term: 0011203...in place of 011203...
·6	**Officers.** Take 1 counter from any *longer* row. [B. Descartes (1953)].	No period found. $\mathcal{G}(10342) = 256$.
·007	**Treblecross.** One-dimensional tic-tac-toe. [WW, 93–94].	No pattern yet found.
·077	**Duplicate Kayles.** Knock down 2 or 3 contiguous skittles [Guy & Smith (1956)].	$p = 24$. Kayles with each nim-value repeated, 00112233114433...
·7777	**Double Kayles.** Take up to 4 beans from a heap; leave rest in at most 2 heaps. [Guy & Smith (1956); WW, 98].	$p = 24$. Kayles with each nim-value g replaced by the pair $2g, 2g+1$ or $2g+1, 2g$ (according to a certain rule), 0123456732897654328945...
·156	See Kenyon (1967).	$p = 349$.
·165	See Austin (1976).	$p = 1550$.
4 · 3	**Lasker's Nim**	0124356879... $p = s = 4$.
·8	(first cousin of) **Triplicate Nim.** Take 1 from heap, rest left in exactly 3 non-empty heaps.	Arithmetically periodic, $p = 3$, saltus = 1. 0000111222333444...
·3F	(F=15) **Kenyon's Game.** Take 1 from heap or take 2 & leave rest in any number of heaps up to 3. [Kenyon (1967)].	$p = 6$, $s = 3$ 0120123453456786789...
·E	(E =14) Take 1, leave rest in just 1, 2 or 3 heaps.	001234153215826514... $\mathcal{G}(246) = 128$. No known pattern.

Table 3. Some sample take-and-break games.

hexadecimal games with $\mathbf{d}_k \le \mathbf{15} = \mathbf{F}$. Such games may be arithmetically periodic. Recently, Anil Gangolli & Thane Plambeck have established the ultimate periodicity of four octal games which were previously unknown:

·**16** has period 149459 (a prime!), the last exceptional value being $\mathcal{G}(105350) = 16$. The game ·**56** has period 144 and last exceptional value $\mathcal{G}(326639) = 26$. The games ·**127** and ·**376** each have period 4 (with cycles of values 4, 7, 2, 1 and 17, 33, 16, 32 respectively) and last exceptional values $\mathcal{G}(46577) = 11$ and $\mathcal{G}(2268247) = 42$.

Grundy's Game [Grundy (1939); WW, 111], split a heap into two *unequal* heaps, continues to defy complete analysis, despite Mike Guy's calculation of the first ten million nim-values. Among these,

$$0, 1, 6, 7, 10, 11, 12, 13, 18, 19, 20, 21, 24, \ldots$$

occur quite rarely. When written in binary, these values contain an even number of ones if you ignore the last digit. These **rare** values form a closed space (the **sparse space**) under nim-addition, while the complement forms the **common coset**:

$$\begin{array}{ccccccccc} \text{rare} & \overset{*}{+} & \text{rare} & = & \text{rare} & = & \text{common} & \overset{*}{+} & \text{common} \\ \text{rare} & \overset{*}{+} & \text{common} & = & \text{common} & = & \text{common} & \overset{*}{+} & \text{rare} \end{array}$$

If the nim-values in a sequence begin to cluster in a suitable common coset, this clustering is likely to persist. In Kayles the rare and common values are **evil** and **odious** numbers respectively, with an even and odd number of ones in their binary expansions. On the other hand, Dawson's Kayles doesn't exhibit this sparse space phenomenon. In Grundy's Game only 1273 rare values have appeared; $\mathcal{G}(82860) = 108$ being the only one in the range $36184 < n \le 10^7$. If the rare values have indeed died out, then Grundy's Game will ultimately be periodic, but the period may be astronomical.

Amongst the comparative chaos, John Conway & Mike Guy have noted a remarkable structure in the nim-values for Grundy's Game, related to the number 59. The probability that $\mathcal{G}(n + d) = \mathcal{G}(n)$ is often as high as $\frac{1}{4}$ if

$$d \quad \text{is } near \quad 59k \quad \text{and} \quad d \equiv k \bmod 3.$$

Examples of these pseudo-periods are 58, 61, 116, 119, 122, 290, 293, 296, 360, 412, 580, 583, 586, 589, 647, 650, 882, 952, 1172, where the last four correspond to $k = 11, 15, 16, 20$.

6 Green Hackenbush

[ONAG, 165–172; WW, 183–190]

This is played on a picture, as in Blue-Red Hackenbush, but now all the edges are grEen, and may be chopped by Either player, making it an impartial game. Every Green Hackenbush picture has a nim-value: for example (Figure 6(c)) the value of a string of 6 green edges is $*6$. It is clear that the six possible moves exactly parallel the six possible moves that you can make from a heap of six beans.

We will see how to evaluate Green Hackenbush trees by the Colon Principle and how to reduce any picture to a forest by the Fusion Principle.

Green Hackenbush trees are examples of the **ordinal sum** $G : H$ which can be defined [WW, 214] for any two games G and H

$$G : H = \{G^L, G : H^L | G^R, G : H^R\}$$

where any move in G annihilates H, while a move in H leaves G unaffected. The **Colon Principle** [WW, 184–185] states that $H \geq K$ implies $G : H \geq G : K$, and, in particular, that $H = K$ implies $G : H = G : K$. That is, $G : H$ depends only on the *value* of H and not on its *form*. It *may* depend on the *form* of G, because there are games $G_1 = G_2$ for which $G_1 : H \neq G_2 : H$.

The Colon Principle applies at branch points of Green Hackenbush trees, allowing us to do nim-addition "up in the air." For example, in Figure 6(a) at a, $*3+*2=*$; at b, $*+*=0$; and at c, $*+*2=*3$, so the value is the same as that of Figure 6(b), where, at d, $*2+*2+*+*4=*5$, and the tree is worth $*6$. Notice the interplay of ordinary addition along strings, with nim-addition at branch points.

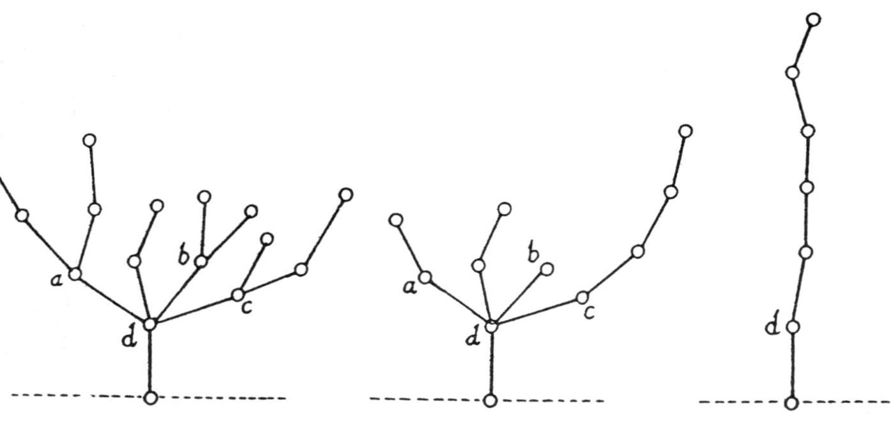

Figure 6(a). Figure 6(b). Figure 6(c).

Figure 6: Transforming a tree into a stalk.

Green Hackenbush pictures involving circuits can be evaluated by the **Fusion Principle** [WW, 186–188]: the value of such a picture is unaltered if you identify the nodes of a circuit. The edges of the circuit then become loops, which may be replaced by twigs (compare Figures 7(b) & 7(c)). Check that the value of Figure 7(a) is $*8$. In this way, every component of a Green Hackenbush picture can be reduced to a tree, and hence to a string, and the strings are combined by nim-addition.

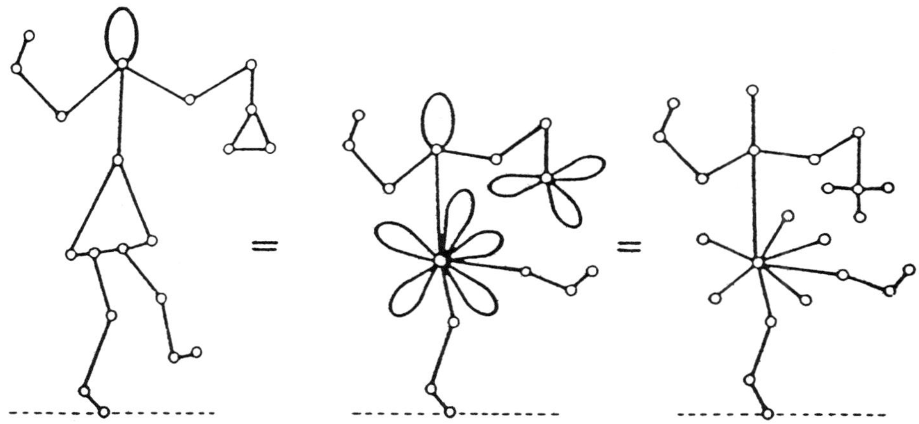

Figure 7(a). Figure 7(b). Figure 7(c).

Figure 7: Girl becomes tree.

7 Welter's Game

[Welter (1952, 1954); Berlekamp (1972); ONAG, 153-165; WW, 472–481]

This is another game whose analysis involves the interplay of nim-addition and ordinary addition. Richard Nowakowski will go into it in more detail than I shall have time for. It is a form of Nim with unequal heaps, but in order to keep track of empty heaps, only one of which is allowed, it's better to play it with coins on a strip of squares, numbered 0, 1, 2, ..., with at most one coin on a square. A move is to shift a coin leftwards to any unoccupied square, possibly passing over other coins. The game ends when the k coins are on the leftmost squares 0, 1, ..., $k-1$. Figure 8 shows a position with $k = 7$.

Figure 8: The position $\{1,2,3,5,8,13,21\}$ in Welter's Game.

To calculate the nim-value, or **Welter function**, $[a|b|c|\ldots]_k$ of the position with k coins on squares a, b, c, ..., first note that for $k = 1$, $[a] = a$, and that for $k = 2$, $[a|b]$ is one less than the nim-sum of a and b: e.g., $[1|3] = 1$, $[5|6] = 2$. For more than two coins, **mate** the pair that is congruent modulo the highest power of two (it doesn't matter that this pair may not be unique). Remove the mated pair and find the best mated pair among the remaining $k - 2$ coins. Continue until all coins are mated, except, when k is odd, for one coin, the **spinster**, s. Then, if (a, b), (c, d), ... are the mates, $[a|b|c|d|\ldots]$ may be calculated as the *nim-sum*

$$[a|b] \overset{*}{+} [c|d] \overset{*}{+} \ldots (\overset{*}{+} [s])$$

where the last term is included just if k is odd.

In Figure 8 the best mates are (5,21), then (1,13), then (2,8), and 3 is the spinster, so the nim-value is

$$\begin{aligned}[1|2|3|5|8|13|21] &= [5|21] \overset{*}{+} [1|13] \overset{*}{+} [2|8] \overset{*}{+} 3 \\ &= 15 \overset{*}{+} 11 \overset{*}{+} 9 \overset{*}{+} 3 \quad = \quad 14.\end{aligned}$$

It turns out that $[a|b|c|d] = 0$ just if the nim-sum $a \overset{*}{+} b \overset{*}{+} c \overset{*}{+} d = 0$, so Welter's Game with four coins can be played with a Nim-like strategy. To play with three coins, imagine a fourth coin on an extra square -1, and add one to each of the numbers labelling the squares while you calculate your move. E.g., {2,5,8} is like {0,3,6,9}, where the winning move would be to {0,3,5,6,}, so, in the three-coin position, move to {2,4,5}.

The mating method makes it easy to calculate the nim-value of a Welter position, but it's not so easy to find the good moves which make the nim-value zero. However, there's a remarkable connexion with **frieze patterns** [Conway & Coxeter (1973); WW, 475–480], which work for nim-addition as well as for multiplication and ordinary addition, and which allow you (or your computer) both to calculate the value of the Welter function and to invert it.

Start with a row of zeros above the Welter position that you want to evaluate, and manufacture a frieze pattern (so called because, when it is extended to the right, it eventually repeats periodically) by completing diamonds

$$\begin{matrix} & b & \\ a & & d \\ & c & \end{matrix} \qquad \text{using the rule} \qquad a \overset{*}{+} d = (b \overset{*}{+} c) + 1,$$

so that $c = [a|d] \overset{*}{+} b$, where the sums are still nim-sums. Lo and behold (Figure 9) the value of the Welter function appears at the foot of the pattern, as follows from a formula on page 159 on ONAG.

0		0		0		0		0		0		0		0
	1		2		3		5		8		13		21	
		2		0		5		12		4		23		
			3		7		13		15		31			
				3		12		13		11				
					9		13		10					
						15		11						
							14							

Figure 9: Calculating the Welter function from a frieze pattern.

If you want to change the value $n = [a|b|c|\dots]$ to some $n' \neq n$, then there are unique $a' \neq a$, $b' \neq b$, $c' \neq c$, ... such that

$$[a'|b|c|\ \dots] = n' = [a|b'|c|\ \dots] = [a|b|c'|\ \dots] = \dots$$

and $[a|b|c|\ \dots] = n$ remains true if we replace any *even* number of a, b, c, ..., n by the corresponding primed letters. This **Even Alteration Theorem** [ONAG, 160–162; WW, 477] may be written

$$\begin{bmatrix} a \mid b \mid c \mid & \\ a' \mid b' \mid c' \mid & \cdots \end{bmatrix} = \begin{matrix} n \\ n' \end{matrix}$$

To find a', b', c', ... corresponding to a given n', continue the bottom row of the frieze pattern, n, n', n, n', n, ... alternately, and then extend the pattern to the right, using the same diamond rule. You will discover that the defining row, a, b, c, ... continues with the answers, a', b', c', ... !

0		0		0		0		0		0		0		0		0		0		0		0		0		0		0
	1		2		3		5		8		13		21		15		0		37		35		10		11		19	
		2		0		5		12		4		23		25		14		36		5		40		0		23		
			3		7		13		15		31		24		25		41		5		15		45		29			
				3		12		13		11		17		25		33		15		12		9		47				
					9		13		10		6		31		46		4		7		11		8					
						15		11		0		9		41		8		13		7		11						
							14		0		14		0		14		0		14		0							

Figure 10: Inverting the Welter function using a frieze pattern.

In Figure 10 we find the good moves in the position {1,2,3,5,8,13,21} by choosing $n' = 0$ and extending the pattern of Figure 9. If you extend it even further to the right, you'll see why it's called a frieze pattern. If you believe the algorithm, and read the second row of Figure 10,

$$\left[\begin{array}{c|c|c|c|c|c|c} 1 & 2 & 3 & 5 & 8 & 13 & 21 \\ 15 & 0 & 37 & 35 & 10 & 11 & 19 \end{array}\right] = \begin{array}{c} 14 \\ 0 \end{array}$$

Check that each move leads to a $\mathcal{P}$-position. Some of the suggested moves, e.g., 1 to 15, 3 to 37, are not legal, but, provided $n' < n$, you'll always find one that is legal, in fact there is always an odd number of legal good moves. Here there are three good moves:

2 to 0, 13 to 11, and 21 to 19.

We can even give you a strategy for the misère form (last player losing) of Welter's Game, if you're willing to learn about **Abacus Positions** [see WW 478–481].

8 Coin-Turning Games

[WW, 429–456]

Several of the impartial games we've already mentioned, and a wide range of new games, can be realized by an idea of Hendrik Lenstra. The $\mathcal{P}$-positions in several of these turn out to correspond to the codewords in some well-known and some not-so-well-known error-correcting codes. To save Vera Pless some time in her lecture, we'll give the rules of some of them here.

Turning Turtles was originally played with turtles, but it's less cruel to play it with a row of coins (Figure 11). A move is to turn a head to a tail, with the additional option of turning at most one other coin, to the left of it. This second coin may go from head to tail, or from tail to head. The game is over when all coins show tails, and the last player wins.

Figure 11: A Turning Turtles position, with coins 3, 4, 6, 7 showing heads.

We leave you to verify that this is a disguise for Nim: if you number the coins 1, 2, 3, ... from the left, then the nim-value of coin n is $*n$ if it's a head, and 0 if it's a tail. The nim-value of a general position is the nim-sum of the nim-values, i.e., the nim-sum of the nim-values of the heads. For example, the good moves in Figure 11 are to turn coin 6 to a tail; or to turn 7 to a tail and 1 to a head; or to turn 4 to a tail and 2 to a head.

Mock Turtles is played in the same way, but a move may turn one, two or three coins, provided the rightmost turned goes from head to tail (this is to make the game satisfy the ending condition). We now number the coins from *zero* (the Mock Turtle) and find the nim-value (or Grundy function), $\mathcal{G}(n)$, of the nth coin, when head up, to be:

n	=	0	1	2	3	4	5	6	7	8	9	10	11	12	13	14	15	16	17	18 ...
$\mathcal{G}(n)$	=	1	2	4	7	8	11	13	14	16	19	21	22	25	26	28	31	32	35	37 ...

These are the **odious numbers** which we met as common values in Kayles.

$$\mathcal{G}(n) = 2n \text{ or } 2n + 1.$$

To find which, write n in binary and append a check digit, 0 or 1, to make an *odd* number of digits 1.

Moebius, **Mogul** and **Moidores** are the corresponding games in which a move turns up to t coins, where $t = 5$, 7 and 9. We consider only odd values of t, because the **Mock Turtle Theorem** gives us the results for even values of t:

> Every nim-value for the $t = 2m + 1$ game
> is an odious number.
> The corresponding value for the $t = 2m$ game
> is got by dropping the final binary digit.

The nim-values for coins 0 to 17 (when head-up) in Moebius are shown in Table 4. The structure of the $\mathcal{P}$-positions in 18-coin Moebius is revealed on replacing the coin numbers by the labels in the third row.

coin #	0	1	2	3	4	5	6	7	8	9	10	11	12	13	14	15	16	17
nim-value	1	2	4	8	16	31	32	64	103	128	171	213	256	301	342	439	475	494
label	∞	1	4	0	-4	-1	5	6	-8	2	-3	-5	8	3	-7	7	-6	-2

Table 4. Eighteen-coin Moebius gives the game its name.

Coins 0 to 5, with labels ∞, 0, ± 1, ± 4, clearly form a $\mathcal{P}$-position (whichever ones you turn over, I'll turn over the rest). Starting from this, or indeed from any $\mathcal{P}$-position, we can find others by operating on the labels with any **Möbius transformation** (modulo 17):

$$x \to \frac{ax+b}{cx+d} \quad \text{with} \quad ad - bc = 1.$$

There are	1 +	102 +	153 +	153 +	102 +	1	$\mathcal{P}$-positions
with respectively	0	6	8	10	12	18	heads.

If we drop the Mock Turtle (at ∞) we have the $t = 4$ game on 17 coins. The $\mathcal{P}$-positions in these two games correspond to the codewords in the [18,9,6] extended quadratic residue code and the [17,9,5] quadratic residue code.

Similarly if we play 24-coin Mogul ($t = 7$, turn up to 7 coins) we find

	1 +	759 +	2576 +	759 +	1	$\mathcal{P}$-positions
with	0	8	12	16	24	heads

coinciding with the 2^{12} codewords of the extended [24,12,8] Golay code. With $t = 6$ and 23 coins the $\mathcal{P}$-positions correspond to the codewords in the perfect [23,12,7] Golay code. Robert Curtis [**1, 2**] has given a pictorial representation of this, the Miracle Octad Generator or "MOG", which also shows the connexion with the Steiner system $S(5, 8, 24)$.

In the **Ruler Game** any number of *contiguous* coins may be turned (with the rightmost always going from head to tail). If the coins are numbered from 1, then the nim-value, $\mathcal{G}(n)$, is the highest power of 2 that divides n.

In **Turnips** (or **Ternups**) a move turns three equally spaced coins. Number the coins from 0 and write n in *ternary* (base 3). Then $\mathcal{G}(n)$ is the kth odious number if the last digit 2 in the ternary expansion is in the kth place from the right, or $\mathcal{G}(n) = 0$ if there is no digit 2 in the ternary expansion of n.

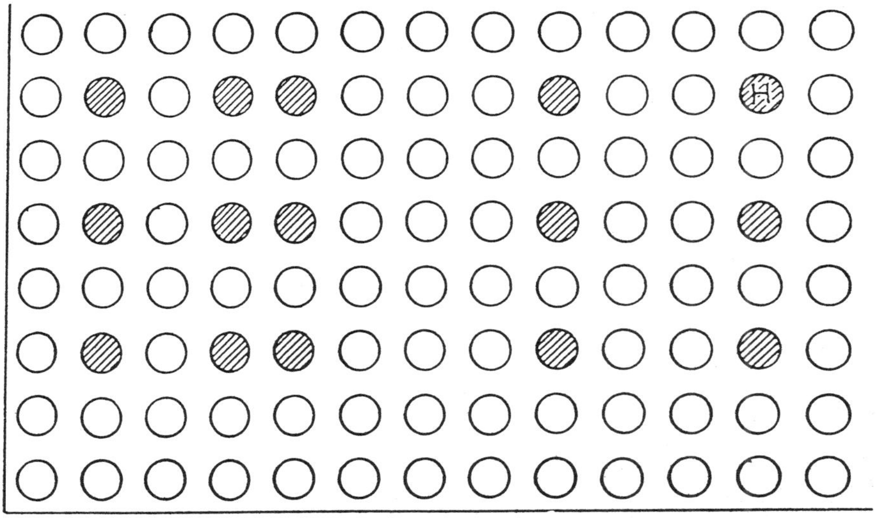

Figure 12: A typical move in Moebius $\times$ Turnips.

There is a plethora of such coin-turning games. They can also be played on a two-dimensional array of coins. For example, we can play the Cartesian product, $A \times B$, of two one-dimensional games A and B, in which a move is to turn all coins with coordinates (a_i, b_j), where $\{a_i\}$ and $\{b_j\}$ are sets of coins constituting legal moves in games A and B respectively. To satisfy the ending condition, the "most northeasterly" coin turned must go from head to tail (Figure 12).

The nim-value of a (head-up) coin in such a game is given by the **Tartan Theorem**:

> The nim-values for the game $A \times B$ are the nim-products of those for A and B:
>
> $$\mathcal{G}_{A+B}(a,b) = \mathcal{G}_A(a) \overset{*}{\times} \mathcal{G}_B(b)$$

where $\mathcal{G}_A(a)$ is the nim-value of coin number a in game A, etc., and $\overset{*}{\times}$ denotes **nim-multiplication**. Nim-multiplication (ONAG, 52-53; Lenstra [1977]) may be defined from the field laws (e.g., associativity and distributivity over nim-addition), together with the rule

$$n \overset{*}{\times} N = n \times N \quad \text{for} \quad n < N$$
$$N \overset{*}{\times} N = 3N/2$$

where N is any **Fermat power** of 2 $(2, 4, 16, \ldots, 2^{2^h}, \ldots)$. For example, $2 \overset{*}{\times} 2 = 3$, because 2 is a Fermat power, while $2 \overset{*}{\times} 3 = 2 \overset{*}{\times} (2 \overset{*}{+} 1) = 3 \overset{*}{+} 2 = 1$. A more complicated example is

$$13 \overset{*}{\times} 7 = (4 \overset{*}{\times} 3 \overset{*}{+} 1) \overset{*}{\times} 7 = (4 \overset{*}{\times} (2 \overset{*}{+} 1)) \overset{*}{\times} (4 \overset{*}{+} 2 \overset{*}{+} 1) \overset{*}{+} 7 =$$

$$4 \overset{*}{\times} 4 \overset{*}{\times} (2 \overset{*}{+} 1) \overset{*}{+} 4 \overset{*}{\times} 2 \overset{*}{\times} (2 \overset{*}{+} 1) \overset{*}{+} 4 \overset{*}{\times} (2 \overset{*}{+} 1) \overset{*}{+} 7 =$$

$$6 \overset{*}{\times} (2 \overset{*}{+} 1) \overset{*}{+} 4 \overset{*}{\times} (3 \overset{*}{+} 2) \overset{*}{+} 4 \overset{*}{\times} 3 \overset{*}{+} 7 = (4 \overset{*}{+} 2) \overset{*}{\times} (2 \overset{*}{+} 1) \overset{*}{+} 4 \overset{*}{\times} 1 \overset{*}{+} 12 \overset{*}{+} 7 =$$

$$4 \overset{*}{\times} 2 \overset{*}{+} 2 \overset{*}{\times} 2 \overset{*}{+} 4 \overset{*}{+} 2 \overset{*}{+} 4 \overset{*}{+} 11 = 8 \overset{*}{+} 3 \overset{*}{+} 9 = 2.$$

The assiduous reader will verify that the nim-cube-roots of 1 are 1, 2 and 3, and the nim-fifth-roots are 1, 8, 10, 13, 14.

9 Misère Nim and an Awful Warning

The power of the Sprague-Grundy theory derives from its reduction of all impartial games to the game of Nim. But particular games may not break up naturally into disjunctive sums, so that much of the force is lost. There are other reasons why it may be difficult or tedious or *hard* in the technical sense, to calculate the nim-value of a position. Also, the theory applies only to normal play.

When Bouton gave his analysis of Nim, he also noted that only a small modification is needed to cover the **misère** version, in which the last player *loses*.

> *Play just as in ordinary Nim until all the heaps, with just one exception, contain a single bean. Then take all the beans from the exceptional heap, or all the beans but one, so as to leave an* odd *number of heaps of size one.*

It's tempting to think (and several people have been tempted to write) that you can play misère impartial games just like normal impartial games until very near the end, when you ...

BUT THAT'S NOT TRUE!

The situation is very complicated. The little that is known in general is given in Chapter 13 of WW. However, an intriguing breakthrough was made recently by William Sibert and John Conway, who have given an analysis of **Misère Kayles**.

REFERENCES

1. Robert T. Curtis, A new combinatorial approach to M_{24}, *Math. Proc. Cambridge Philos. Soc.,* **79**(1976) 25–42; MR **53**#3098.

2. Robert T. Curtis, The maximal subgroups of M_{24}, *Math. Proc. Cambridge Philos. Soc.,* **81**(1977) 185–192; MR **55**#12807.

3. Richard K. Guy, Fair Game, COMAP Math. Exploration Series, Arlington MA, 1989.

Proceedings of Symposia in Applied Mathematics
Volume **43**, 1991

More Ways of Combining Games

John Horton Conway

1 Introduction

These notes, and those on Numbers and Games, were largely prepared by Richard Guy to whom I am more grateful than I can say. I quote from his paper [**1**]:

> "who has written a good deal of this paper (though not all of this sentence), ..."

2 A Dozen Ways for Simultaneous Plays

[Smith (1966); ONAG, 173–187]

So far we've combined games by their (disjunctive) **sum**, and usually assumed normal play, with the last player winning in the last component to end. But there are other ways to play several games simultaneously. When it's your turn to move, instead of playing in just ONE component, you might move in ALL of them, or in SOME (maybe all, but at least one). We'll call these alternative combinations **joins** or **unions**, respectively, to distinguish them from sums.

How do games end? For sums and unions, it's natural to continue so long as there's a component with a legal move. For joins, it's more natural to stop as soon as play ends in *any* component, since it's no longer possible to move in *every* component. However, it's possible to consider the opposite state of affairs in each case. These possibilities are

The paper is in final form and no version of it will be submitted for publication elsewhere.
1980 *Mathematics Subject Classification* (1985 *Revision*). Primary 90D05.

	name	compound	symbol and pronunciation	moves made in	finish is determined when	analysis
$\mathcal{P}$	Proper sum	disjunctive	$G + H + K + \ldots$ Gplus$H$plus...	ONE	ALL (long)	*Plain nim-value
$\mathcal{Q}$	Quick sum	*diminished* disjunctive	$G \oplus H \oplus K \oplus \ldots$ Gpless$H$pless...	ONE	FIRST (short)	*Queer, or foreclosed nim-value
$\mathcal{R}$	Rapid join	conjunctive	$G \wedge H \wedge K \wedge \ldots$ $G$and$H$and...	ALL	FIRST (short)	Remoteness
$\mathcal{S}$	Slow join	*continued* conjunctive	$G \triangle H \triangle K \triangle \ldots$ Galso$H$also...	ALL	ALL (long)	Suspense number
$\mathcal{T}$	Tardy union	selective	$G \vee H \vee K \vee \ldots$ Gor$H$or...	SOME	ALL (long)	Tally (Toll & Timer)
$\mathcal{U}$	Urgent union	*severed* selective	$G \triangledown H \triangledown K \triangledown \ldots$ Gur$H$ur...	SOME	FIRST (short)	Unrestricted tally
				component(s)	component(s) is(are) finished	

Table 1. Six ways to combine and end games.
*In the first two cases the function described only solves the theory for *impartial* games.

summarized in Table 1, with mnemonic hints, using the initial letters of the names of the games, for the methods of analysis.

And who is the winner? In **normal play** the last person to play wins: you lose if you're unable to move. But in **misère play** the last player to move *loses*. These two outcomes, combined with the six possibilities of Table 1, provide a dozen methods of conducting simultaneous play. We will comment briefly on a few of these. But we have no theory for misère unions of partizan games, and misère sums, even for quite simple impartial games, quickly get beyond our grasp.

For each of these combinatorial operations, say "$\odot$", we ask:

just what must we know about G, H, K, ...
in order to tell who wins $G \odot H \odot K \odot \ldots$?

The answer is their equivalence classes under the natural equivalence relation defined by:

$G \approx H$ just if $G \odot X$ and $H \odot X$
have the same outcome for all X.

We usually specify "$\approx$" by defining a suitable function f in such a way that we have $G \approx H$ precisely when $f(G) = f(H)$. These functions are suggested by the entries in the 'analysis' column of Table 1.

*In the first two cases the function described only solves the theory for *impartial* games.

In order to analyze joins, we need to know something about

3 The Remoteness Function

[WW, 259–266]

Steinhaus (1925) gave an early analysis of impartial games, using what C.A.B. Smith (1966) has since called the **remoteness function**. This is useful for games where the idea of sum doesn't arise naturally, or doesn't arise at all, e.g., in **joins** of games in which moves must simultaneously be made in *all* components. It can apply to partizan games, where we will define *Left* and *Right* remotenesses, and in other places where we may want to amend the rules: for example, in misère play, where the last player *loses*, or where the ending condition doesn't hold (so that some remotenesses may be infinite).

Intuitively, the remoteness is the number of turns that the game lasts when the winner is trying to win as quickly as possible, while the loser tries to postpone defeat as long as possible. In normal play the last player wins, so she wants to move to a terminal position (remoteness 0), and, more generally, to a position of *even* remoteness, forcing her opponent always to move to positions of odd remoteness. Terminal positions have remoteness 0; positions with an option of remoteness 0 have remoteness 1; positions *all* of whose options have remoteness 1, are of remoteness 2; and so on:

Rules for calculating the remoteness

If there's an option of *even* remoteness:	1 + least even remoteness	$\mathcal{N}$-position
If *all* options have *odd* remoteness:	1 + greatest odd remoteness	$\mathcal{P}$-position
If there are *no* options:	0	terminal position

The $\mathcal{P}$-positions are those of *even* remoteness; and the $\mathcal{N}$-positions those of *odd* remoteness. If the game doesn't satisfy the ending condition, it may not always be possible to continue assigning remotenesses according to these rules. The remaining positions are $\mathcal{O}$-positions (open positions, for which neither player has a winning strategy), to which we assign *infinite* remoteness.

As an example of a game where there is only one component, and the ending condition doesn't hold, so that nim-values are of no use to us, we will consider Epstein's (1967) **Put-or-Take-a-Square** game [WW, 484–486, 501–502]. This is played with a single heap of beans. A move is to add or take away the largest perfect square number of beans in the heap. The object is to take the last bean, so the empty heap has remoteness 0, and a positive perfect square number of beans has remoteness 1.

Figure 1 shows some heap sizes with legal moves indicated by arrows.

We can next assign remoteness 2 to heaps of 5 and 20, because *both* options have remoteness 1. Then 11, 14, 21, 30, 41, 54 have remoteness 3, since in each case there's an option of remoteness 2 (and no option of remoteness 0). A heap of 29 has remoteness 4: because both options (4 & 54) have odd remoteness, and the *larger* is 3.

Each of 2 and 3 is an option of the other, and *sensible* players won't choose the other options, 1 and 4, because these have (odd) remoteness

rem	$\mathcal{P}$-positions	$\mathcal{N}$-positions	rem
0	0		
		1,4,9,16,25,36,49,64,81,100,121, 144,169,196,225,256,...	1
2	5,20,45,80,145,...		
		11,14,21,30,41,44,54,69,86,105,120, 126,141,149,164,174,189,216,291, ...	3
4	29,101,116,135,165,236,...		
		52,71,84,208,254,284,296,...	5
6	257,...		
		136,436,...	7
8	477,...		
		252,342,...	9
10	173,...		
		92,...	11

Table 2. Remotenesses of heap sizes in Epstein's Put-or-Take-a-Square Game.

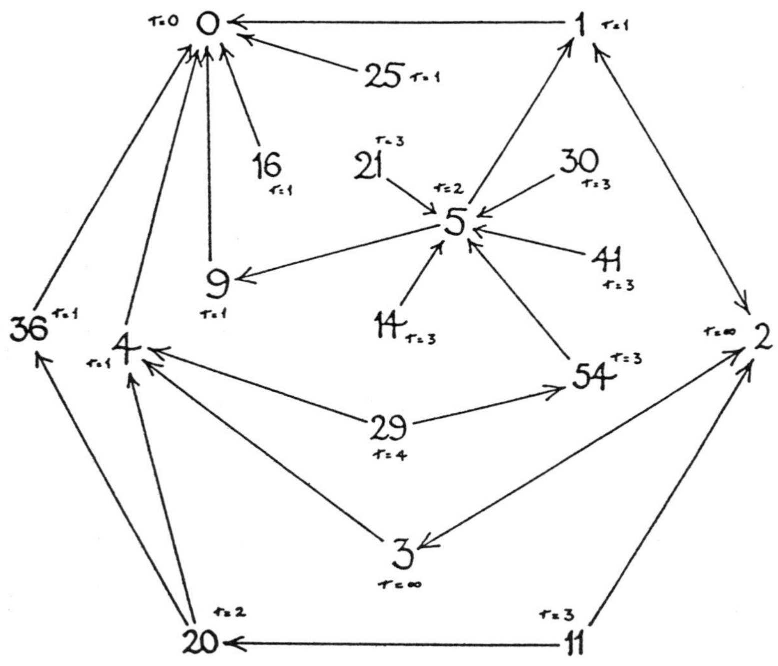

Figure 1: Assigning remotenesses in Epstein's Game.

1 and lose (immediately). So best play goes 2, 3, 2, 3, 2, ... and the remotenesses of 2 and 3 are both infinite. Table 2 shows a few $\mathcal{P}$- and $\mathcal{N}$-positions, but the great majority of positions,

2, 3, 6–8, 10, 12, 13, 15, 17–19, 22–24, 26–28, 31–35, 37–40,
42, 43, 46–48, 50, 51, 53, 55–63, 65–68, 70, 72, ...

are $\mathcal{O}$-positions with infinite remoteness. A more extended table is given on page 486 of WW.

4 Joins

[WW, 257–278]

We can analyze (rapid) joins, partizan or impartial, normal or misère, using Steinhaus's remoteness function. If the game is partizan, calculate the Left and Right remotenesses separately, respectively using the remotenesses of the Right and Left options. For misère play, reverse the parity of the normal remoteness rules (Table 3).

In each component, to calculate the

LEFT NORMAL	RIGHT NORMAL	LEFT MISÈRE	RIGHT MISÈRE

remoteness, take it as ZERO if there's no option for

LEFT	RIGHT	LEFT	RIGHT

Otherwise ADD ONE to the LEAST

EVEN RIGHT	EVEN LEFT	ODD RIGHT	ODD LEFT

remoteness of any

LEFT	RIGHT	LEFT	RIGHT

option which has such a remoteness, else ADD ONE to the GREATEST

ODD RIGHT	ODD LEFT	EVEN RIGHT	EVEN LEFT

remoteness of any

LEFT	RIGHT	LEFT	RIGHT

option which has such a remoteness.

Table 3. Remoteness rules for rapid joins.

For impartial joins, omit all references to Left and Right, giving a single normal (or misère) remoteness function. Since a rapid join of games ends when the *first* component ends, the remoteness (left or right or impartial, normal or misère) of the RAPID JOIN is the

LEAST REMOTENESS of any component.

To win a rapid join, move to a position in which your *opponent's* remoteness is

EVEN in NORMAL PLAY
ODD in MISÈRE PLAY

To play SLOW JOINS, in which play continues in the remaining components even though some have already ended, use the **suspense number** [WW, 266–272], which has the opposite philosophy to that of remoteness: use cat-and-mouse tactics; if you're winning, spin the game out as long as possible; if you're losing, aim to get it over quickly. Convert Table 3 to give **suspense rules** by interchanging GREATEST and LEAST throughout. The suspense of a SLOW JOIN is then the

GREATEST SUSPENSE of any component,

and, to win a slow join, again move to a position where your *opponent's* suspense number is EVEN or ODD according as play is NORMAL or MISÈRE.

We illustrate these ideas with an analysis of several variants of the game, **All the King's Horses**, taken, with kind permission of Academic Press and the authors, from Chapter 9 of WW. This is played as a join: each player moves *every* horse in one or other of the two ways indicated in Figure 2. There can be arbitrarily many horses on a square and they are *all* moved by *both* players. A player is unable to move if there is any one horse he can't move. Under normal play he then loses, but under misère play he wins. Left and Right remotenesses are calculated as shown in Figure 3. Table 4 shows the Left and Right remotenesses (a) in normal play, and (b) in misère play.

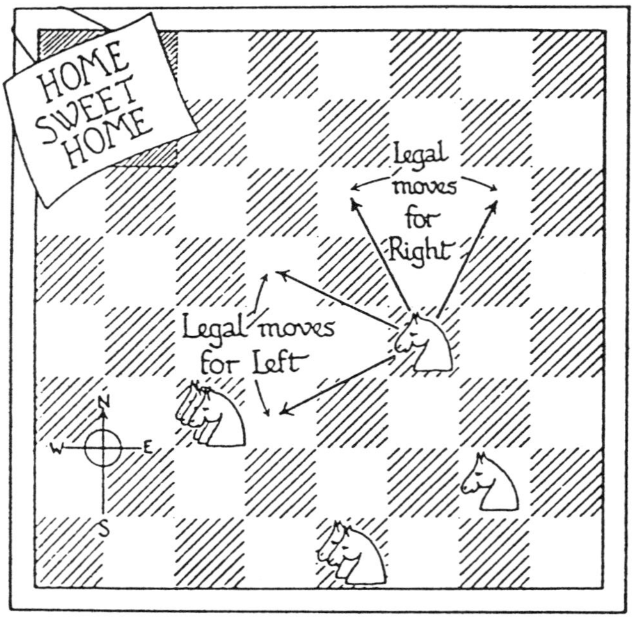

Figure 2: How horses head for home.

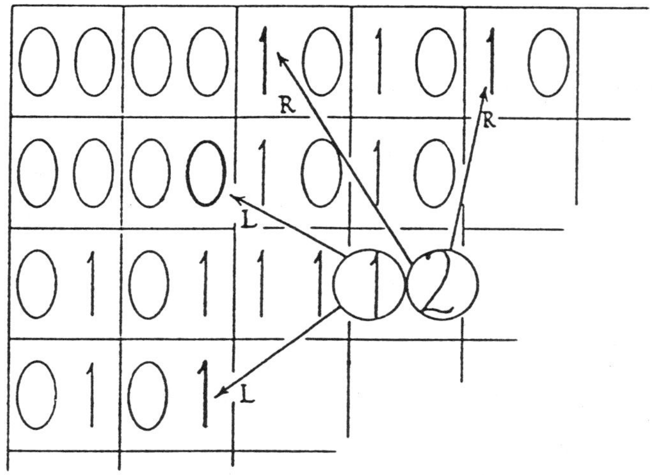

Figure 3: How remote is a horse?

00	00	10	10	10	10	10	10
00	00	10	10	10	10	10	10
01	01	11	12	12	12	12	12
01	01	21	22	22	32	32	32
01	01	21	22	22	32	32	32
01	01	21	23	23	33	34	34
01	01	21	23	23	43	44	44
01	01	21	23	23	43	44	44

00	00	10	10	10	10	10	10
00	00	20	30	30	30	30	30
01	02	22	42	52	52	52	52
01	03	24	44	64	74	74	74
01	03	25	46	66	86	96	96
01	03	25	47	68	88	X8	x8
01	03	25	47	69	8X	XX	TX
01	03	25	47	69	8x	XT	xx

(a) First horse stuck loses. (b) First horse stuck wins [X=10, x=11, T=12].

Table 4. Left and Right remotenesses in (a) normal, and (b) misère, play.

If we play All The King's Horses as a **slow join**, so that the *last* horse to reach home determines the outcome, we must allow a player to pass for a horse he can't move: the game ends when *all* the horses reach home. Best play is guided by **suspense numbers**, shown in Table 5.

We can also play this version as an *impartial game*: each player moves every horse whenever possible, using any of the four moves shown in Figure 2. Play is guided by a single suspense number, as in Table 6.

00	00	12	12	34	34	56	56
00	00	12	12	34	34	56	56
21	21	11	12	32	34	54	56
21	21	21	22	22	34	34	56
43	43	23	22	22	34	34	56
43	43	43	43	43	33	34	54
65	65	45	43	43	43	44	44
65	65	65	65	65	45	44	44

(a) Last horse home wins.

00	00	12	12	45	34	56	56
00	00	23	12	34	34	67	56
21	32	22	22	44	36	56	56
21	21	22	44	34	44	56	77
54	43	44	43	44	56	66	56
43	43	63	44	65	66	58	66
65	76	65	65	66	85	66	77
65	65	65	77	65	66	77	66

(b) Last horse home loses.

Table 5. Left and Right suspense numbers for a slow join: (a) normal play, (b) misère play.

0	0	1	1	2	2	3	3
0	0	1	1	2	2	3	3
1	1	1	3	3	3	3	3
1	1	3	3	3	3	5	5
2	2	3	3	4	4	5	5
2	2	3	3	4	4	5	5
3	3	3	5	5	5	5	5
3	3	3	5	5	5	5	6

(a) Last horse home wins.

0	0	1	1	3	2	3	3
0	0	2	1	2	2	4	3
1	2	2	4	2	4	4	4
1	1	4	3	2	3	6	5
3	2	2	2	5	4	3	4
2	2	4	3	4	3	6	5
3	4	4	6	3	6	5	6
3	3	4	5	4	5	6	7

(b) Last horse home loses.

Table 6. Impartial suspense numbers (a) in normal, and (b) in misère, play.

Another variant, a compromise between the rapid join and the slow join, is to allow a player to pass for a horse he can't move, provided his opponent *can* move it. Then, the first horse *home* determines the outcome.

In Figure 4, the misère remotenesses are being calculated. Left can move to a position of Right remoteness 3, so his remoteness is $3 + 1 = 4$. Right has no move, but can pass to the same position with Left to move, so the right remoteness is $4 + 1 = 5$. Table 7 shows the remotenesses for this variant, and Table 8 gives the (single) remotenesses when it is played impartially, each player having up to four possible moves for each horse.

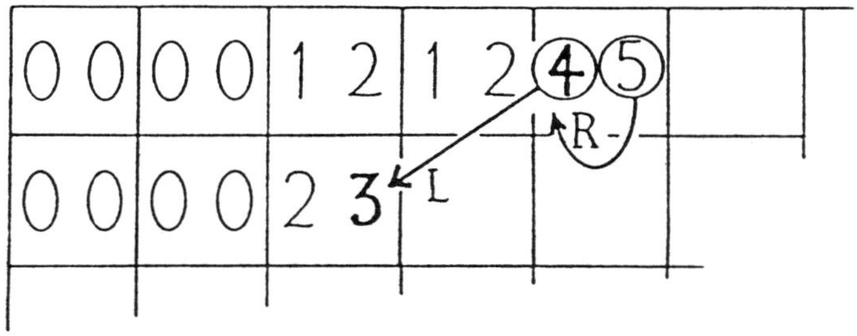

Figure 4: Right's a bit more remote because he's stuck.

00	00	12	12	34	34	56	56
00	00	12	12	34	34	56	56
21	21	11	14	34	36	56	56
21	21	41	44	44	56	56	76
43	43	43	44	44	56	56	76
43	43	63	65	65	55	58	76
65	65	65	65	65	85	88	86
65	65	65	67	67	67	68	66

(a) First horse home wins.

00	00	12	12	45	56	56	56
00	00	23	34	34	34	67	78
21	32	22	42	42	56	56	56
21	43	24	44	54	64	74	77
54	43	24	45	66	76	86	76
65	43	65	46	67	66	98	X8
65	76	65	47	68	89	88	x9
65	87	65	77	67	8X	9x	XX

(b) First horse home loses.

Table 7. Left and Right remotenesses for a not so rapid join: (a) with normal play, (b) with misère play [X=10, x=11].

0	0	1	1	2	2	3	3
0	0	1	1	2	2	3	3
1	1	1	1	3	3	3	3
1	1	1	3	3	3	3	5
2	2	3	3	4	4	5	5
2	2	3	3	4	4	5	5
3	3	3	3	5	5	5	5
3	3	3	5	5	5	5	6

(a) First horse home wins.

0	0	1	1	3	4	3	3
0	0	2	3	2	2	4	5
1	2	2	2	2	4	4	4
1	3	2	3	4	5	6	5
3	2	2	4	5	4	5	6
4	2	4	5	4	7	6	7
3	4	4	6	5	6	7	6
3	5	4	5	6	7	6	7

(b) First horse home loses.

Table 8. Impartial remotenesses for the not so rapid join: (a) with normal play, (b) with misère play.

5 Unions

[WW, 279–306]

Before we analyze unions, in which you're allowed to move in any positive number of components, remind yourself of the distinction between hot and cold games. Domineering is an example of a hot game and Blue-Red Hackenbush is a cold game. For our present purposes the following simple example, in which Left and Right each have a single option which is a number, will suffice.

If $x < y$, then the game $\{x|y\}$ is **cold**,
i.e., a number, in fact the simplest number between x and y

If $x > y$, then the game $\{x|y\}$ is **hot**.

If $x = y$, then the game $\{x|y\}$ is **tepid**,
in fact it is $x + *$, written $x*$.

The best strategy in playing a union of several components is to move in *all* the *hot* ones, and in *none* of the cold. The first phase of the play is like a *join* of the hot components. Players will want to continue so long as there's a hot component left, so it's a *slow join*. When all the components are cold (i.e., numbers), players will be reluctant to play and will move in only one component, the one where least harm is done. The union becomes like an ordinary sum.

The selective theory of unions combines the disjunctive theory of sums with the conjunctive theory of slow joins.

We indicate the state of play by a pair of (Left and Right) tallies [WW, 280–306]. A **tally** consists of a toll; and a timer, written as a subscript. If Left (or Right) starts, the Left (or Right) **toll** is the (numerical) value that the component will acquire when the hot phase is over. The Left (or

Right) **timer** is the number of moves that the hot phase lasts. Think of the timers as suspense numbers (though care is needed in their calculation for *tepid* games: see later).

To find the Left (or Right) tally for the union of several components, *add* the Left (or Right) tolls and take the *greatest* of the Left (or Right) timers.

The best *Left* option in a component is one with the *greatest Right* toll, and, among those, the one with *largest even* timer (or the *least* if they're all *odd*). The best *Right* option is one with *least Left* toll, and, among those, make the same choice of timer (largest even, else least odd).

To find the tallies for a component, suppose that the best Left option has Right tally x_a, and the best Right option has Left tally y_b.

If $x > y$, the game is *hot*, and the tallies are $x_{a+1}y_{b+1}$.

If $x < y$, the game has become *cold* with tallies z_0z_0, where z is the *simplest* number (with earliest birthday) in the interval

$x < z < y$ if a and b are both even	*even* timers *exclude* their tolls,
$x \leq z \leq y$ if a and b are both odd,	*odd* timers *admit* their tolls,
$x < z \leq y$ if a is even and b is odd,	as candidates for
$x \leq z < y$ if a is odd and b is even.	the simplest number.

If $x = y$, the game is *tepid*. *Try* $x_{a+1}y_{b+1}$: this is correct if $a+1$, $b+1$ are both odd, but if they are both even, replace them by 0 (the game has become cold). If just one of $a+1$, $b+1$ is even, increase the other (if necessary) by just enough to make it a larger odd number.

There's not room here to prove these rules, but we'll illustrate them with the game **One for Left, Two for Right, Free for All**, a modification of **·007**, (**Treblecross**; knock down three contiguous skittles from a row): made *partizan* (and more tidy) by allowing Left to knock down rows with just one skittle, and Right to knock down rows of just two: and made *selective* by allowing players to attack as many rows as they like in a single move.

In contrast to Treblecross, we can give a complete analysis of this game, though you're forgiven if you don't immediately see the pattern in Table 9.

Row of	$3k+1$	$3k+2$	$3k+3$	Row of	$3k+1$	$3k+2$	$3k+3$
$k=0$	1	-1	$0_1 0_1$	$k=7$	$1_3 1_5$	$2_4{-}1_4$	$0_5 0_5$
$k=1$	$1_1 1_1$	$2_1{-}1_1$	$0_1 0_1$	$k=8$	$1_3 1_5$	$2_4{-}1_6$	$0_5 0_4$
$k=2$	$1_2 2_1$	$2_2{-}1_2$	0	$k=9$	$1_4 1_5$	$2_4{-}1_6$	$0_7 0_4$
$k=3$	$1_2 1_3$	$2_2{-}1_2$	$0_3 0_3$	$k=10$	$1_5 1_5$	$2_4{-}1_6$	$0_7 0_5$
$k=4$	$1_1 1_3$	$2_4{-}1_1$	$0_3 0_3$	$k=11$	$1_5 1_5$	$2_4{-}1_8$	$0_7 0_4$
$k=5$	$1_2 1_3$	$2_4{-}1_4$	$0_3 0_2$	$k=12$	$1_5 1_5$	$2_6{-}1_8$	$0_9 0_4$
$k=6$	$1_4 1_5$	$2_4{-}1_4$	$0_5 0_3$	$k=13$	$1_5 1_5$	$2_6{-}1_8$	$0_9 0_6$

Table 9. Tallies for One for Left, Two for Right, Free for All.

We calculate the tallies for a row of 16 skittles, which a legal move may reduce to rows of

$0_{\vee}13$	$1_{\vee}12$	$2_{\vee}11$	$3_{\vee}10$	$4_{\vee}9$	$5_{\vee}8$	$6_{\vee}7$

skittles, which have tallies (earlier entries in Table 9)

$0_{\vee}1_1 1_3$	$1_{\vee}0_3 0_3$	$-1_{\vee}2_2{-}1_2$	$0_1 0_{1\vee}1_2 1_3$	$1_1 1_{1\vee}0$	$2_1{-}1_{1\vee}2_2{-}1_2$	$0_1 0_{1\vee}1_2{-}2_1$
i.e., $1_1 1_3$	$1_3 1_3$	$1_2{-}2_2$	$1_2 1_3$	$1_1 1_1$	$4_2{-}2_2$	$1_2{-}2_1$
1_3	1_3		1_3	1_1		

are the tallies with best Right toll from Left's point of view. They have no even timer: the least odd is 1. The best Left tolls for Right are

1_1	1_3	1_2	1_2	1_1		1_2

and the greatest even timer is 2. Our rules show that $\{\dots 1_1 | 1_2 \dots\}$ is tepid and tell us to try $1_2 1_3$. This is correct: although there's an even timer, the other is odd and greater.

The general pattern for the tolls is easy to see: The only exceptions are the Left toll for two skittles and the Right toll for seven. The pattern for the timers is complicated, and further obscured by 22 exceptions. Timers increase at three different rates, one twice another, the third only as the logarithm (to base 2) of the first. If l_i, r_i are the Left and Right timers for a row of $3k+i$ skittles ($i = 1, 2, 3$), then

$$
\begin{aligned}
l_1 &= 2\lfloor (k+3)/6 \rfloor + 1 \quad (k \ge 10) & r_1 &= 2\lfloor \log_2(k+2) \rfloor - 1 \quad (k \ge 3)\\
l_2 &= 2\lfloor \log_2(k+4) \rfloor - 2 \quad (k \ge 2) & r_2 &= 2\lfloor (k+1)/3 \rfloor \quad (k \ge 5)\\
l_3 &= 2\lfloor (k+3)/3 \rfloor - 1 \quad (k \ge 3) & r_3 &= 2\lfloor (k+5)/6 \rfloor \quad (k \ge 11)
\end{aligned}
$$

We don't have a misère theory for tardy unions, but urgent unions can be dealt with, both for normal and misère play, by introducing two new kinds of move: an **overriding** move which *wins* immediately; and a **suiciding** one which *loses* immediately: and we allow infinite tolls for such options. For precise details see WW, 292–306.

REFERENCE

1. Richard K. Guy, How to factor a number, Proc. 5th Manitoba Conf. Numer. Math., Winnipeg, 1975, 49–89, esp. p. 73.

Proceedings of Symposia in Applied Mathematics
Volume **43**, 1991

Introductory Overview of Mathematical Go Endgames

ELWYN BERLEKAMP

INTRODUCTION AND OVERVIEW OF VARIOUS DIALECTS OF THE RULES OF GO

This paper does not assume any prior knowledge of the game of Go, but it assumes familiarity with mathematical game theory as presented in the previous papers of this session or in part 1 of *Winning Ways.*

The game of Go originated in China about 3,000 or 4,000 years ago. It remains very popular today, especially in China, Japan, Korea, Taiwan, and Hong Kong. By almost any measure, the popularity of Go in Japan exceeds the combined popularity of bridge and chess in the United States. Most Japanese newspapers have regular Go columns. There are hundreds of local clubs throughout the country, many of which have a resident professional somewhat like golf clubs in the US. These professional Go players receive income from teaching amateurs and also from winning tournament prizes. There are eight major well-funded title matches in Japan alone, and many more in neighboring countries.

Go is a two-person perfect-information game of skill. It is played on a graph printed on a board. Although Go can be played on any graph, the conventional ones have the nodes arranged in a square array, with the branches being the unit horizontal and vertical lines.

The two players are called Black and White. We define Left as Black and Right as White. Each of these players starts with a pot containing a plentiful supply of stones of his own color. Black plays first, and turns then alternate. At each turn, the player normally places one stone of his color onto an empty node of the board. Each turn is called a *move.* For example, after 6 moves, on an initially-empty board, there will be 3 white stones and 3 black stones on the board. [1]

The strategic objective of the game is to enclose territory and to capture opposing stones.

[1] In a game between two weak players, it is conceivable and legally possible for one to play on the edge and be captured within the first six moves. In this absurd scenario, there would be fewer than 6 stones on the board after the sixth move.

The final version of this paper will be submitted for publication elsewhere.
1980 *Mathematics Subject Classification* (1985 *Revision*). Primary 90D05.

There have been numerous variations or "dialects" of the rules, which have been popular at different times and different places. Even today, Chinese rules and Japanese rules differ in several major respects, and, even within the last decade, there have, at times, been multiple, conflicting variations of the rules played concurrently within the same country. However, many of the differences in rules concern only subtle situations which occur rarely in practice. Many of the great games played in many major title matches are "universal" in the sense that the reader would find it an excellent contest, with the same strategy and the same winner, even if he assumed that the game had been played under a different variation of the rules. Nevertheless, a rare position in w[1] ich the dialects of rules yield different outcomes does occasionally occur.

As a personal philosophy, I think that this situation should be viewed as a challenge to applied mathematicians. Although we are certainly entitled to our own opinions about the relative elegance and of various definitions of the rules, as a practical matter I do not think it productive to lobby Go rules committees. Instead, I think a more fruitful goal is to "mathematize" each of the major dialects of Go. Corresponding to each of the popular dialects, we can seek a tractable set of rules for a closely related mathematical game. I have already made considerable progress in this direction. I have also defined another dialect called "Basic Mathematical Go." This is an interesting but somewhat coarse approximation to all versions of popular Go.

In many interesting positions, including the one shown in the 9x9 full-board problem in figure 1, the winning moves are the same for all major dialects of the rules, including Japanese, Chinese, and Basic Mathematical.

All of these results will appear in a forthcoming book, *Mathematical Go Endgames*. In the present paper, I shall attempt only to state and explain a few results about Basic Mathematical Go.

The successes of the mathematical approach to the study of Go arise from an implicit decomposition theorem, which allows one to express the value of the entire board precisely as the sum of the values of various disjoint portions of the board. Thus, much of the computational work is devoted to obtaining precise evaluations of relatively small partial-board positions. Many larger partial board endgame positions can be found in the Go literature. However, the viewpoint is quite different. Most existing Go books treat partial board problems as if they were the only active area on the board. They typically assume that Black and White will play alternately, and only within the part of the board shown in the problem. Their "solutions" typically yield only a winner and a local integer-valued score, quantities which are woefully insufficient to evaluate the entire board when it is composed of the sum of several such partial-board problems.

Here are the rules of Basic Mathematical Go.

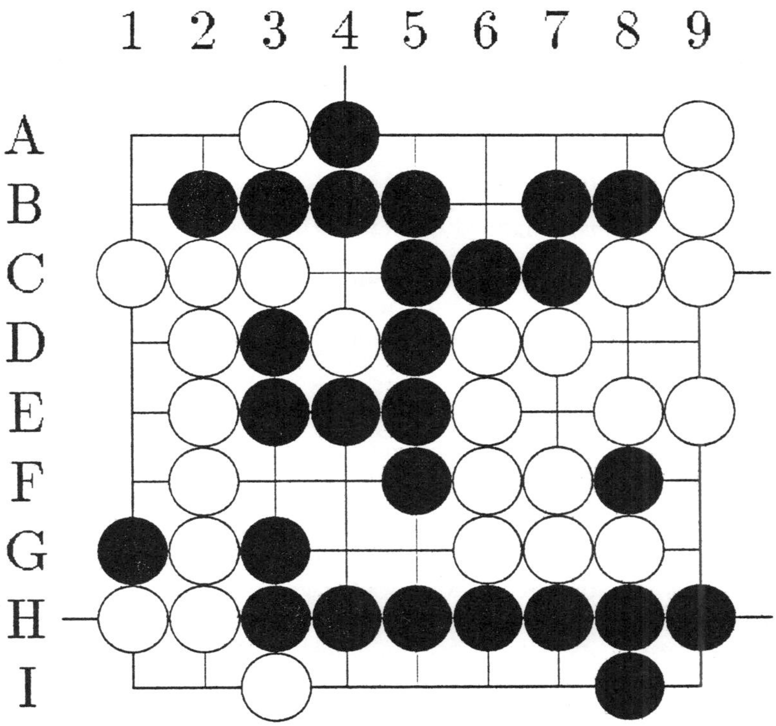

FIG. 1. *White to move*

RULES OF BASIC MATHEMATICAL GO:

Captives & Suicides. Once played, a stone on the board can never be moved to another node. However, under certain conditions it may be removed from the board and placed in a pile of captured stones. These conditions will be specified herewith.

Stones of the same color which occupy nodes connected by horizontal or vertical lines on the board as said to form a *connected group*. Any unoccupied nodes adjacent to this group are said to be *liberties* of the group. The same unoccupied node can be a liberty for more than one group.

If the just-played stone removes the sole liberty of a group of the opposite color, then that group is said to be *captured* and all of the stones belonging to that group are removed from the board. In that case, the move is said to have been a *capturing move*. If the move was not a capturing move, but leaves a group of its own color with no liberties, then the move is said to be a *suicidal move*.

All suicidal moves are banned in basic mathematical Go.

In figure 1, white has 27 legal moves. There are 28 empty nodes, but I9 is suicidal.

Repetitious Positions. In mathematical Go, there are no bans on moves which lead into repetitious loops of positions. It is possible that after 2n moves, the board position is exactly the same as it was before, except that there are now n more captives of each color. If both players decide to continue making moves which repeat the same loop of previously played positions, the outcome of the game is declared to be a DRAW. [2] [3]

Passing. In *all* mathematical dialects of the rules, *passing is banned.*

Special Option To Return A Captive. Instead of playing a stone onto the board, a player may instead exercise the option to complete his move by returning a captive of the opponent's color to the pot.

Outcome. When a player, at his turn, is unable to make a legal move, the game ends and the player unable to move loses.

This can happen only if there are no remaining captives of the opposing color, and if all moves onto the board are suicidal.

Immortality By Convention. Mathematical Go is very concerned with the study of partial board game positions, or more simply, "games". Typically, each of these games contains a region of only a few vacant nodes, which are surrounded by occupied nodes. When investigating the play within such a region, we assume that the stones on its boundary are *immortal.* It is convenient to view lines running away from the stones on the boundary of a partial board problem as lines to a special "node at infinity". Since neither player is permitted to play on

---

[2] In popular Go, these loops are described by various names, depending on the number of positions in the loop, as follows:
$2n = 2$ "Ko"
$2n = 4$ "Eternal Life"
$2n = 6$ "Triple Ko"

Popular rules typically ban moves which lead into repetitious loops of 2 positions (Kos), but there are no corresponding bans on moves which lead to repetitious loops of 4, 6, or more positions.

The situation is made more difficult for Westerners to comprehend by the fact that ties are also possible in some dialects of popular Go, yet commonly used Go tournament rules specify very different consequences for games which end in ties and games which end in draws. Unlike western chess tournaments, where a "tie" due to stalemate and a "draw" due to perpetual check are both treated as equivalent outcomes, Go tournament rules typically treat draws as "no result". This may be analogous to forcing the game to be replayed, or to treating it as if BOTH players had lost!

[3] Some Go players may initially overlook the *close* correspondence between this rule and the implicit conventions of popular partial-board problems. In each case, there are three possible outcomes with the same relative desirability:

Popular Partial Board Problem	Basic Mathematical Go
1. Win without relying on a ko	1. Win
2. Win by relying on a ko	2. Draw
3. Lose	3. Lose

The correspondence is strengthened when we pursue upper and lower mathematical bounds by treating draws as wins for Black or as wins for White. These bounds are equivalent to the partial board problem's assumption that a specified player has an excess number of big ko threats in some unshown area of the board.

Completely accurate treatment of kos is attained only by mathematically cumbersome dialects of the rules, beyond the scope of the present paper.

the node at infinity, a line to infinity ensures a permanent liberty, which means immortality. [4]

Earned Immortality. Modern versions of Basic Mathematical Go also contain a rule which specifies a condition under which groups of stones can acquire immortality in the course of the play. Since novice Go players typically require some experience to understand this rule, we omit it from this exposition. Instead, we use our mathematical version of the rules of ancient Chinese Go, which omits this rule.

This completes our statement of the rules of "Basic Mathematical Go" (ancient Chinese dialect).

We now review and reconsider some of the elementary results of mathematical game theory. There are four basic concepts: game, sum, negative, and equality. The last notion, equality, is very fruitfully defined as the concurrence of two weaker conditions: less-equal and greater-equal.

The first two definitions remain virtually unchanged. However, since many loopy games have no negatives, the notion of negative has to be augmented by the more general notion of a "complement," and we will need new definitions of equality and inequality.

DEFINITIONS AND RESULTS

1. Definition of an Abstract game. As in Winning Ways, a general mathematical game is defined as

$$G \stackrel{\text{def}}{=} \{G^L \mid G^R\}$$

The definition of a game is a "top-down" recursion. To complete the definition of G, we must also provide definitions for G^L and G^R, and for all of their followers, and their followers' followers, etc. The set of games consisting of G and their followers and their followers' followers, etc., is called the *positions* of G.

As we shall see in some subsequent examples, Loopy positions are now allowed. In general, all of the games in a repetitious loop must be defined concurrently.

2. Definition of the sum of games. As in Winning Ways, the sum of two games, G and H, is the game G+H wherein Black can make a Left move on G leaving H alone, or he can make a Left move on H leaving G alone. In symbols,

$$(G+H)^L \stackrel{\text{def}}{=} G^L + H \,,\, G + H^L$$

White has a similar set of choices, and the sum is defined as

[4] The composer of the 9x9 problem in figure 1 has assigned immortality to several groups, via "lines to infinity" from A4, C9, H1, and H9. These lines are critical to the solution via the ancient Chinese version of "Basic Mathematical Go" rules. In more refined dialects of the rules, provisions for earned immortality make these lines superfluous.

$$(G+H) \stackrel{\text{def}}{=} \{G^L + H\ ,\ G + H^L \mid G^R + H\ ,\ G + H^R\}$$

NOTES: This addition is immediately seen to be *associative* and *commutative*. It has an *identity*: $0 = \{\ \mid\ \}$ is immediately seen to satisfy this requirement:

$$\text{For all } G,\ G + 0 = 0 + G = G$$

Example 1:

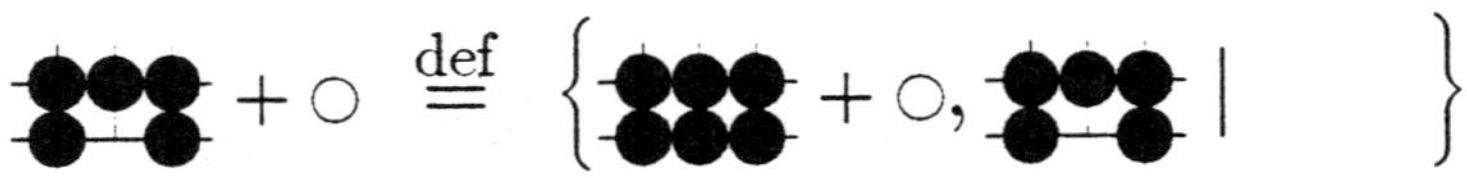

In this simple case, White has no legal move, but Black has a choice. He can either play on the first summand by filling in the vacant node or he can play on the second summand by returning the captive. In symbols,

$$1 + 1 \stackrel{\text{def}}{=} \{0 + 1, 1 + 0 \mid \quad \} = 2$$

Example 2: Definitions of $\textcircled{K}$ *and* $\overline{\textcircled{K}}$*:* Here is a pair of loopy positions which occurs so frequently in late-stage Go endgames that we introduce special symbols for them:

$\textcircled{K} \stackrel{\text{def}}{=}$

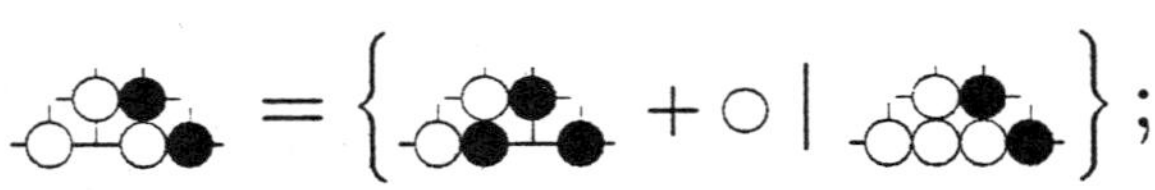

$\overline{\textcircled{K}} \stackrel{\text{def}}{=}$

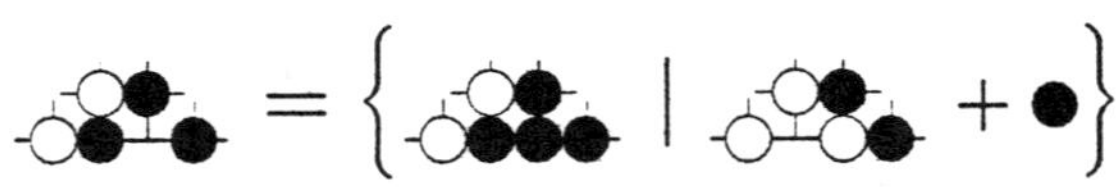

Abstractly, $\textcircled{K}$ and $\overline{\textcircled{K}}$ are defined by these equations:

$$\textcircled{K} = \{1 + \overline{\textcircled{K}} \mid 0\}; \qquad \overline{\textcircled{K}} = \{0 \mid \textcircled{K} - 1\}$$

3a. Definition of the COMPLEMENT:

$$\bar{G} \stackrel{\text{def}}{=} \{\overline{G^R} \mid \overline{G^L}\}$$

THEOREM: *2nd player can win-or-draw on game plus its complement*

Either $G + \bar{G} = 0$

or

Either player, going second, can draw on $G + \bar{G}$

3b. Definition of the NEGATIVE:

If $G + \bar{G} = 0$, then $-G \stackrel{\text{def}}{=} \bar{G}$

If 1st Player can draw on $G + \bar{G}$, then G has no negative

Game trees and Game graphs. We sometimes show top-down "game trees" as the graph whose nodes are the positions of a game, connected by branches showing legal moves. A branch sloping downward to the left represents a legal move for Left; a branch sloping downward to the right represents a legal move for Right. For example,

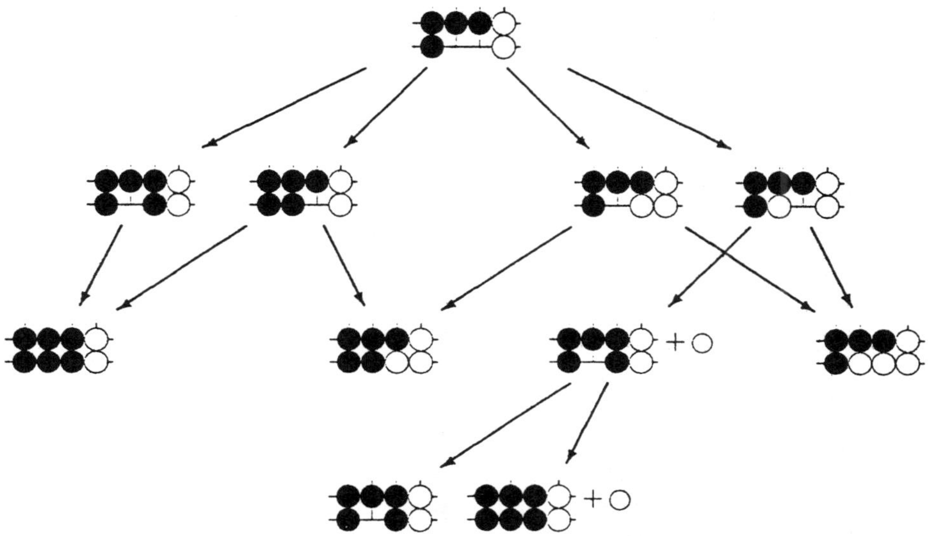

As in this example, from a botanical point of view, the tree is upside down. Its "leaves" are the terminal nodes at the bottom. For brevity, in this example I have treated some integers as leaves, even though, strictly speaking, 0 is the only true "leaf."

The game "tree" corresponding to $\textcircled{K}$ and $\overline{\textcircled{K}}$ is as follows:

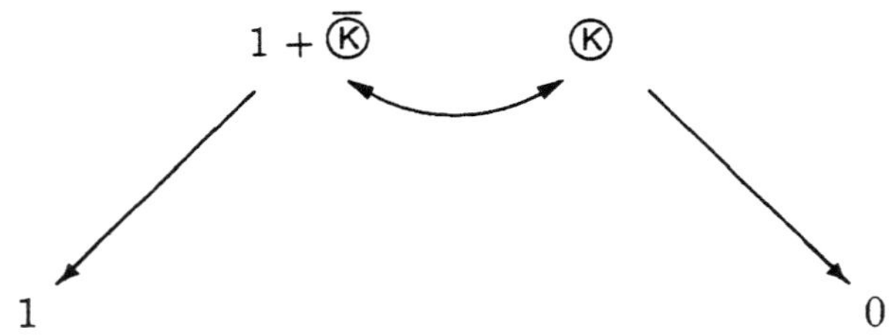

Actually, this graph of the game $\textcircled{K}$ is not a tree; it looks more like a suspension bridge. The "suspension cable" indicates the two-cycle loop between $\textcircled{K}$ and $1 + \overline{\textcircled{K}}$. As this cable should suggest, $1 + \overline{\textcircled{K}}$ is a left follower of $\textcircled{K}$ while $\textcircled{K}$ is a right follower of $1 + \overline{\textcircled{K}}$.

Having illustrated the basic definitions of a "game," "sum," and "negative," we are now ready to state some further definitions and a few elementary results. The need to include games with loopy positions requires some departures from the axioms of Winning Ways. The problem is that many loopy games have no negatives, and since, as a minimum, we surely still want to have every game equal to itself, we cannot define a pair of games to be equal only if their difference equals zero. However, the following slightly more general approach still yields the classical pair of reduction theorems concerning dominated and reversible options, and this allows us to reduce an arbitrary partial-board Go position to a canonical form. In many cases, including some of the examples that we present below, the canonical form is unique and free of loops, even though the original game to which the canonical form is equal may have loopy positions. When the canonical form is loop-free, we can apply all the traditional theorems of Winning Ways directly. In other cases, which are beyond the scope of the present paper, additional theorems are necessary.

We now state the new definitions and a mixture of new and old results:

DEFINITION: Outcomes

$$L_{outcome}(G) \stackrel{\text{def}}{=} \text{winner of well-played } G \text{ if Left plays first}$$

$$R_{outcome}(G) \stackrel{\text{def}}{=} \text{winner of well-played } G \text{ if Right plays first}$$

Both outcomes have a range of three values: {Right, Draw, Left}.

$$outcomes(G) \stackrel{\text{def}}{=} [L_{outcome}(G), R_{outcome}(G)]$$

This is a 2-dimensional vector, whose range has 9 possible values.

DEFINITION: Equality of games

$$G = H \text{ iff for all games X, } outcomes(G + X) = outcomes(H + X)$$

DEFINITION: Partial Ordering of Outcomes.

Adapting our usual perspective ("Left" ~ "positive"), we define a transitive ordering on $L_{outcome}$ and $R_{outcome}$ as follows:

$$\text{Right} < \text{Draw} < \text{Left}$$

We then define a partial ordering of outcomes:

$$Outcomes(G) \leq Outcomes(H) \text{ iff}$$

$$L_{outcome}(G) \leq L_{outcome}(H)$$

$$and$$

$$R_{outcome}(G) \leq R_{outcome}(H)$$

DEFINITION: Partial Ordering of Games

$$G \leq H \text{ iff for all games X, } outcomes(G + X) \leq outcomes(H + X)$$

NOTE 1. We say $H \geq G$ iff $G \leq H$.
NOTE 2. Observe that $G = H$ iff $G \leq H$ and $G \geq H$.

THEOREM: *Transitivity*
If $A \leq B$ and $B \leq C$, then $A \leq C$.

THEOREM: *Additivity*
If $A \leq B$ and $C \leq D$, then $A + C \leq B + D$.

THEOREM: *Game* ≥ 0
$G \geq 0$ iff $R_{outcome}(G)$ =Left
or
$G \geq 0$ iff Left, going second, can win on G.

THEOREM: *Game* ≤ 0
$G \leq 0$ iff $L_{outcome}(G)$ =Right.
or
$G \leq 0$ iff Right, going second, can win on G.

THEOREM: *Game* $= 0$
$G = 0$ iff either player, going 2nd, can win on G.

SIMPLIFICATION THEOREM: *Left's Dominated Options can be Deleted*

Let $G \stackrel{\text{def}}{=} \{A, B, \ldots, C \mid \alpha, \ldots, \beta\}$
Let $H \stackrel{\text{def}}{=} \{B, \ldots, C \mid \alpha, \ldots, \beta\}$
If $A \leq B$, then $G = H$.

SIMPLIFICATION THEOREM: *Left's Reversible Moves can be Bypassed*
Let $G \stackrel{\text{def}}{=} \{A, B, \ldots, C \mid \alpha, \ldots, \beta\}$
Suppose A has a Right follower $A^R = F \stackrel{\text{def}}{=} \{I, J, \ldots, K \mid \gamma, \ldots, \omega\}$
If $F \leq G$, then $G = H \stackrel{\text{def}}{=} \{I, J, \ldots, K, B, \ldots, C \mid \alpha, \ldots, \beta\}$

NOTE: The term "reversible" refers to the values of the games. Be careful to avoid confusion with "repetitious positions," which refers not to the values but to the connectivity among the followers of a loopy game. When we are playing under a dialect of the rules (such as Basic Mathematical Go) which has no ban on immediate ko recaptures that complete 2-cycle repetitious loops in the game graph, then the immediate recapture is both repetitious *and* reversible. However, many reversible moves lead to quite different physical game positions that happen to have the same value, and other reversible moves lead to game positions that are strictly *better* for the reversing player than the position from which the reversible move was made.

SIMPLIFICATION THEOREM: *Right's Dominated Options can be Deleted*
Let $G \stackrel{\text{def}}{=} \{A, \ldots, B \mid \alpha, \beta, \ldots, \gamma\}$
Let $H \stackrel{\text{def}}{=} \{A, \ldots, B \mid \quad \beta, \ldots, \gamma\}$
If $\alpha \geq \beta$, then $G = H$.

SIMPLIFICATION THEOREM: *Right's Reversible Moves can be Bypassed*
Let $G \stackrel{\text{def}}{=} \{A, \ldots, B \mid \alpha, \beta, \ldots, \gamma\}$
Suppose α has a Left follower $\alpha^L = F \stackrel{\text{def}}{=} \{I, J, \ldots, K \mid \epsilon, \ldots, \omega\}$
If $F \geq G$, then $G = H \stackrel{\text{def}}{=} \{A, \ldots, B \mid \epsilon, \ldots, \omega, \beta, \ldots, \gamma\}$

DEFINITION: Canonical Form
A game G is said to be expressed in a *canonical form* iff G and all of its positions are expressed in such a way that there are no dominated options nor any reversible options.

NOTE: According to this version of the definition of a canonical form, a game with a ko has no canonical form, since the ko is reversible to itself. A laxer definition of canonical form would appear to be needed to include games with ko's. However, even with this seemingly strict definition, it is still not yet known whether games with complicated repetitious positions might not be equal to each of several nonidentical canonical forms.

THEOREM: *For games with negatives, the canonical form is unique*
If G and H are both in canonical form, and if either G or H has a negative, and if G = H, then G and H are identical.

THEOREM: *Constructive comparisons*
If $-H$ exists or if $-G$ exists, then $G \geq H$ iff Left, going 2nd, can win on $G - H$.

REMARK: Current status of uniqueness of canonical form in general.
In any class of abstract games within which draws are impossible, every game has a negative. The theory of such games was well-developed in Winning Ways. The present generalization of that theory, which includes games that might have some drawn positions, is very recent. Potentially unique canonical forms for games with draws are not yet understood. It may be that there are some additional simplification theorems for loopy games, and that the application of such theorems will ensure uniqueness of a canonical form in general. It may also be that the canonical form defined above is already unique; neither a proof nor any counterexample is yet known.

EXAMPLE OF A REDUCTION TO CANONICAL FORM

We next present a detailed example of a game with a simple but nonobvious canonical form:

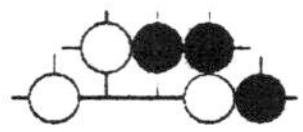

Altogether, this game has 17 different legal on-the-board positions. Here they are, defined in bottom-up order, starting from the true leaves (which are "zero" positions that have no legal followers) and working back towards to the original game which is the root of the game "tree."

 $= \{\ \mid\ \} = 0$

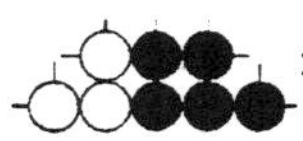 $= \{\ \mid\ \} = 0$

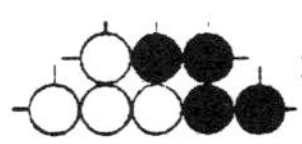 $= \{\ \mid\ \} = 0$

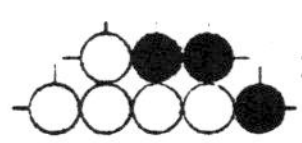 $= \{\ \mid\ \} = 0$

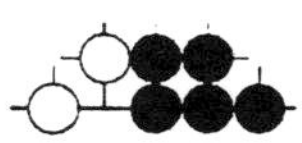 $= \{$ 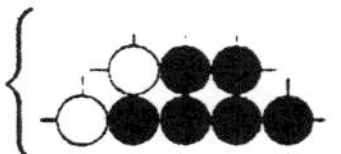 $\mid$ 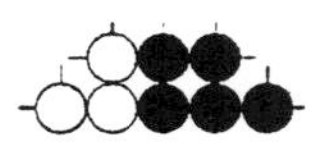 $\} = *$

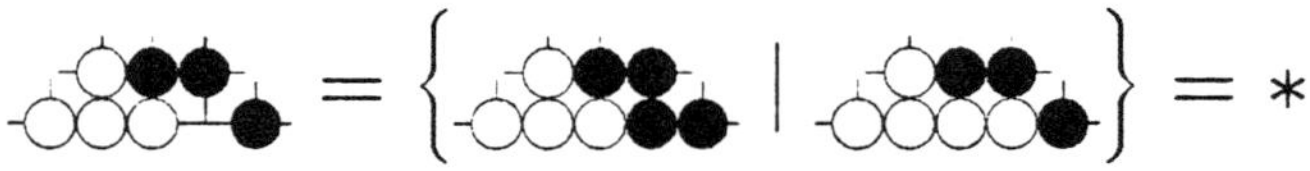
= *

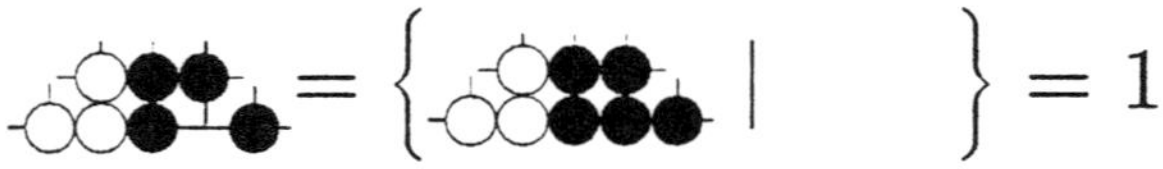
= 1

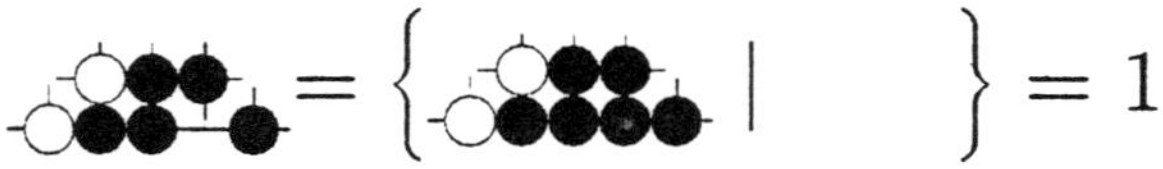
= 1

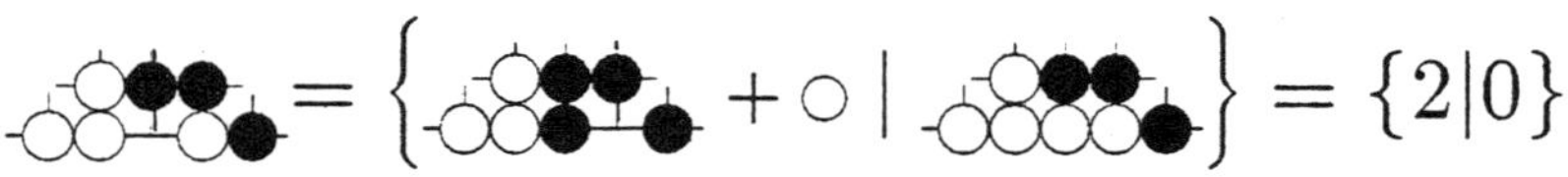
= {2|0}

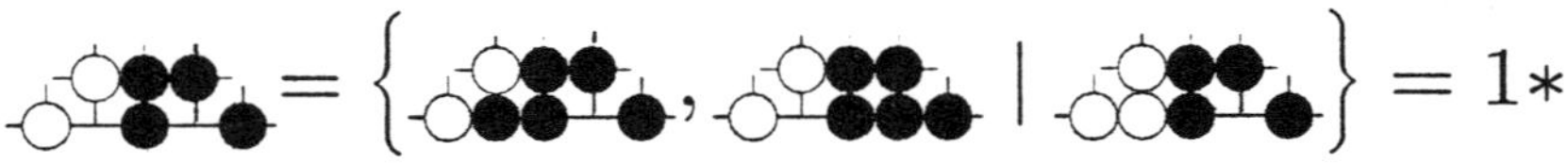
= 1*

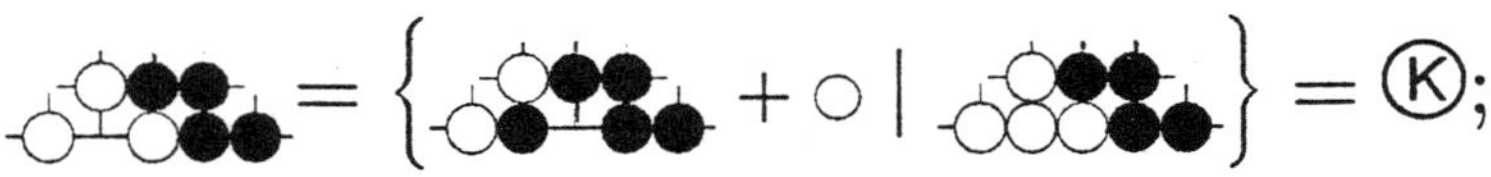
= Ⓚ;

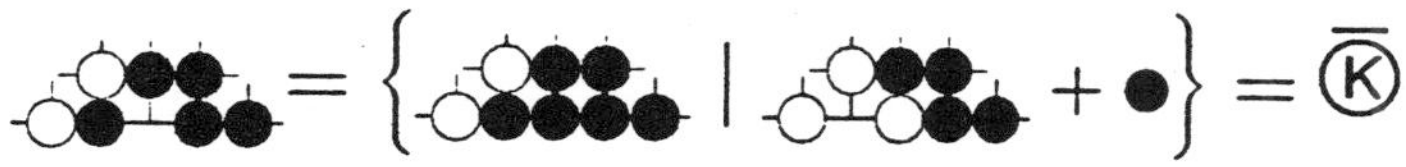

We next come to a set of four positions that must all be defined together:

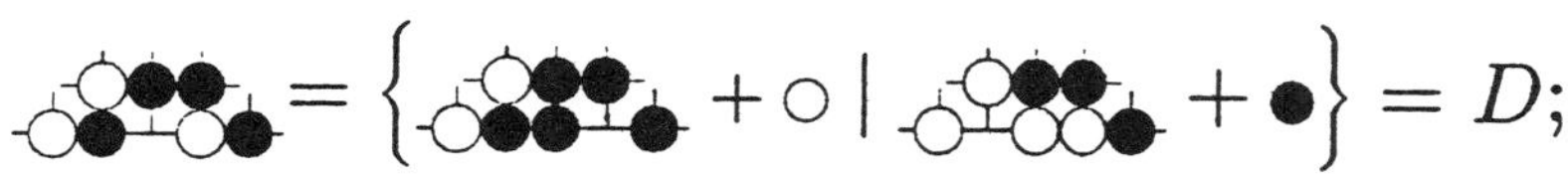

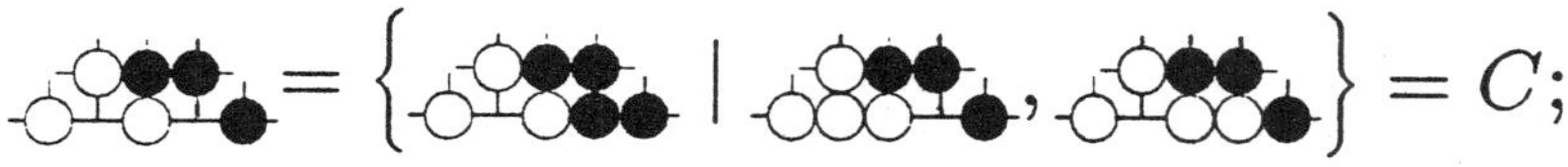

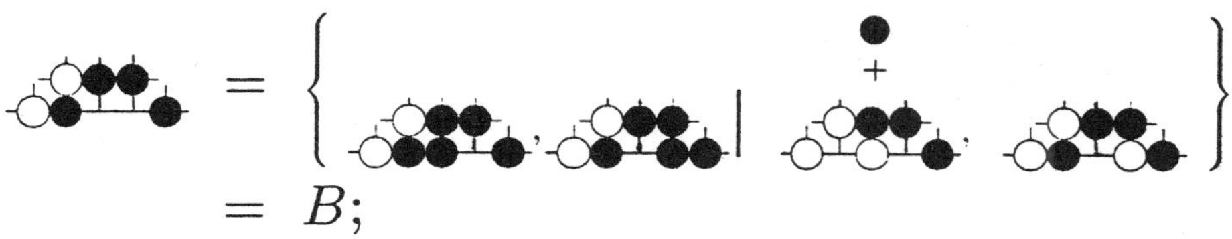

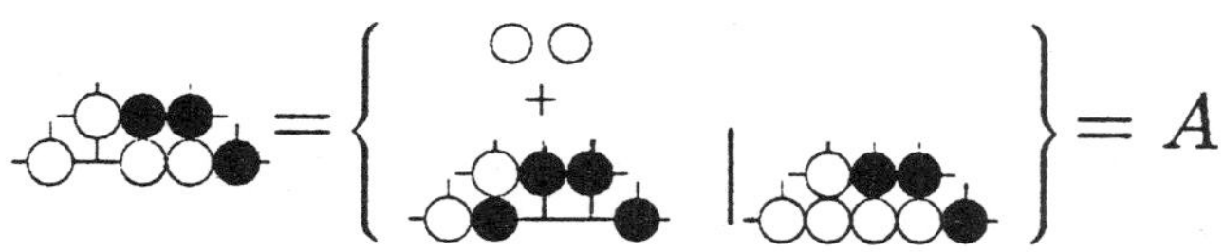

and finally, we arrive at the root of our game tree:

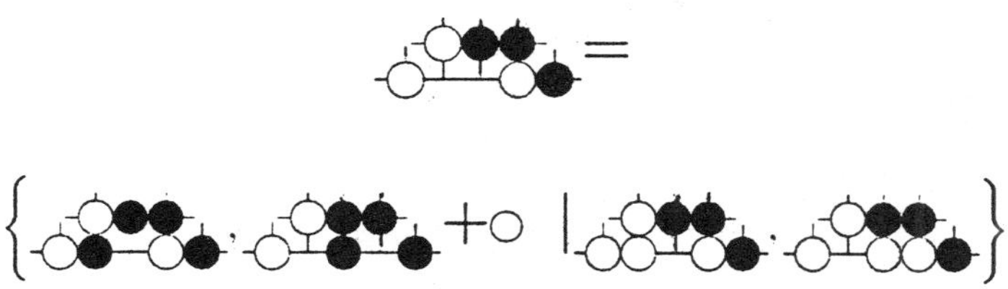

This completes the bottom-up definition of

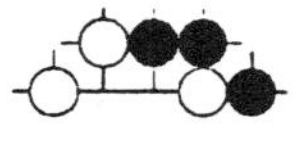

We next compute its canonical form.

The immediate followers of this game included 4 legal positions which were all defined together. These positions can be represented abstractly by the letters A, B, C, and D, which satisfy the relations

$$A = \{2 + B \mid 0\}; B = \{1, \overline{\textcircled{K}} \mid \text{-}1 + C, D\}; C = \{\textcircled{K} \mid *, A\}; D = \{2 \mid \text{-}1 + A\}$$

where

$D =$ 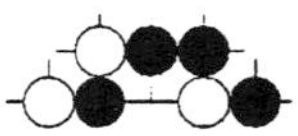

$C =$ 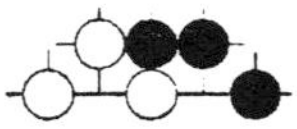

$B =$ 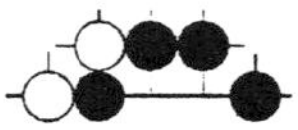

$A =$ 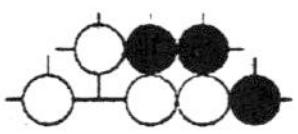

To reduce these games to canonical form, we begin by hunting for dominated options. Comparing the two left options of B, we find that $1 \geq \overline{\textcircled{K}}$, or equivalently, $\overline{\textcircled{K}} - 1 \leq 0$. This is equivalent to the assertion that Right (=White), playing second, can win this game:

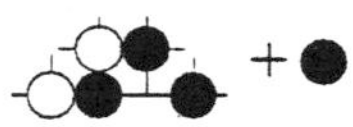

Similarly, comparing the two Right options of B, we find that $-1 + C \leq D$, or equivalently, $0 \leq D + 1 + \bar{C}$. This follows from the assertion that Left (=Black), playing second, can win this game:

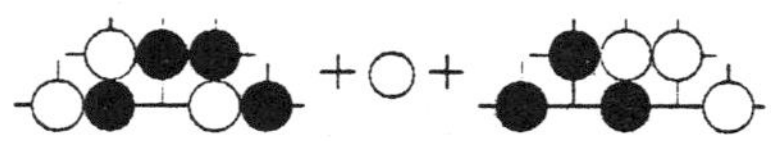

Several different lines of play all need to be verified:
Case 1:

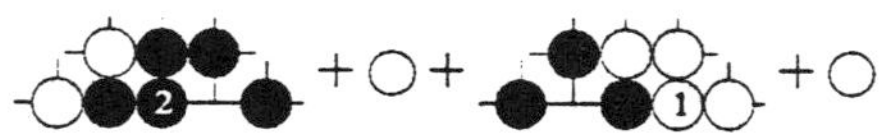

Case 2:

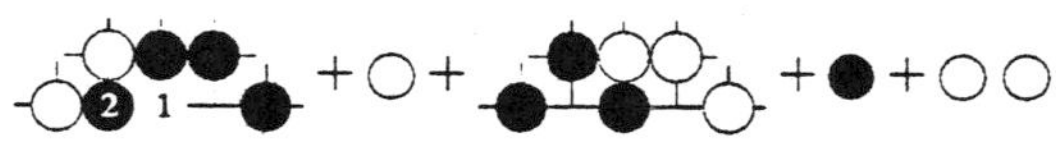

where the black stone is captured on move 1 and the two white stones are captured on move 2.

There are then several continuations:
Case 2a:

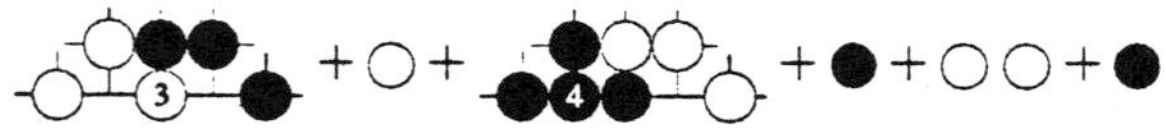

Case 2b:

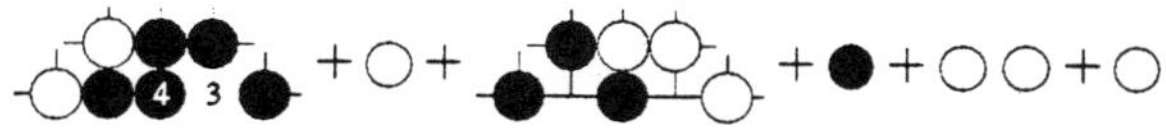

Case 2c:

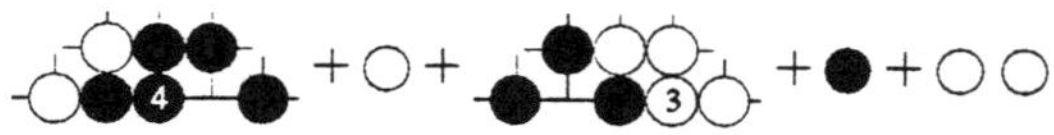

In each of these cases, every possible 5th move by White still leaves some empty node on the board where Black can play his 6th move in such a way that it is connected to immortal Black stones. Thereafter, Black can simply return captives until White runs out of legal moves.

Hence we have shown that

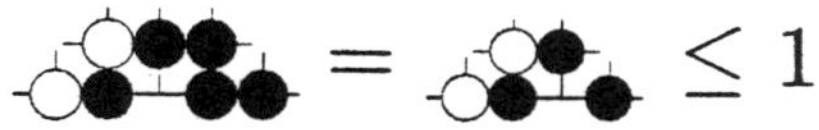

and

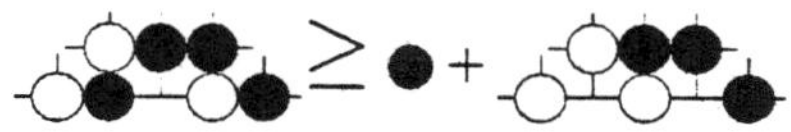

Whence we can delete the dominated options to obtain

$B =$ 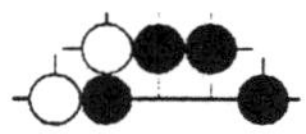

$=$ 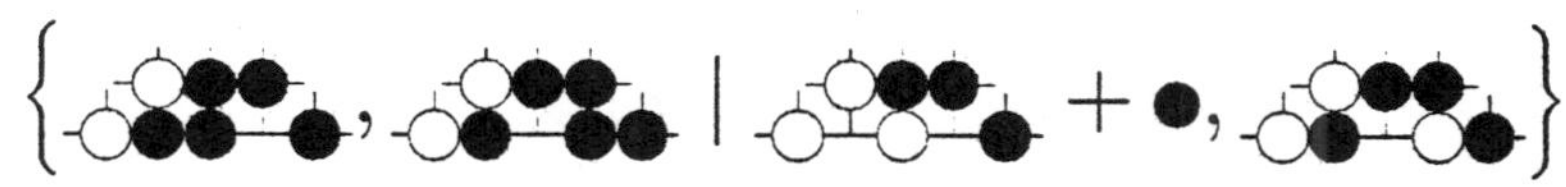

$=$ 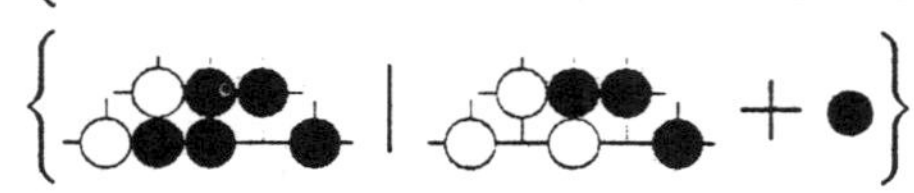

$$= \{1 \mid C - 1\}$$

Similarly, comparing the two Right options of C, we find that $A \geq *$, whence we can delete C's dominated option to obtain

$C =$ 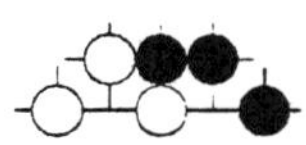

$=$ 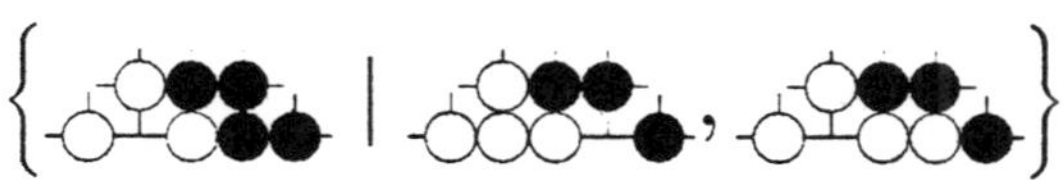

$=$ 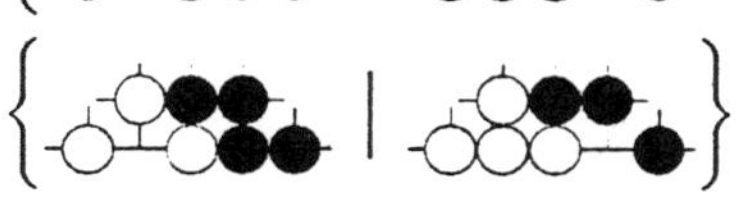

$= \{$ Ⓚ $\mid * \}$

We next search for reversible options. It happens that $C^{RL} = C^{LR} = 0$. Hence, either player, playing second, can win on C. This means that $C = 0$. More specifically, since $0 \le C$, Left's move from C to Ⓚ reverses through 0 to 0^L, which is the empty set, and hence,

$$C = \{\ | *\}$$

Similarly, since $0 \ge C$, Right's move from C to $*$ reverses through 0 to 0^R, which is also the empty set, and hence,

$$C = \{Ⓚ \mid \} = \{|\} = 0$$

This then allows us to simplify the expression for B from

$$B = \{1 \mid \text{-}1 + C\} \text{ to}$$
$$B = \{1 \mid \text{-}1\}$$

Inserting this simplified expression for B into A yields

$$A = \{2 + B \mid 0\} = \{\{3 \mid 1\} \mid 0\}$$

Inserting this simplified expression for A into D yields

$$D = \{2 \mid A - 1\} = \{2 \mid \{\{2 \mid 0\} \mid \text{-}1\}\}$$

Checking for reversible options, we compare D with $D^{RL} = \{2 \mid 0\}$, and find that indeed, $D \le D^{RL}$. We can therefore reverse Right's move from D to $\{\{2 \mid 0\} \mid \text{-}1\}$ through $\{2 \mid 0\}$ to 0, and obtain

$$D = \{2 \mid 0\}$$

We now have A, B, C, and D in canonical form as follows:

$D =$ 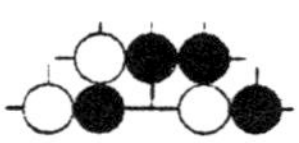 $= \{2 \mid 0\}$

$C =$ 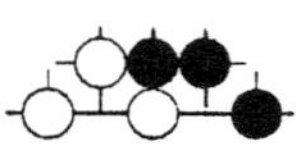 $= 0$

$B =$ 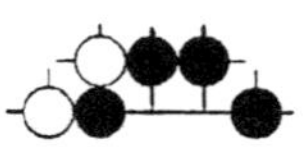 $= \{1 \mid -1\}$

$A =$ 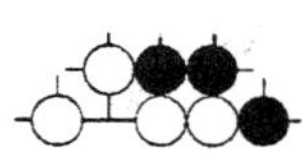 $= \{\{3 \mid 1\} \mid 0\}$

We then consider the original game,

$G =$ 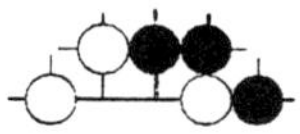

$= \{$ 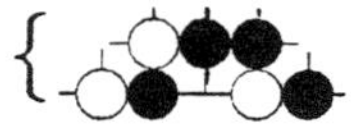 ,  | 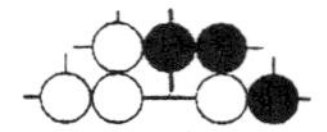 , 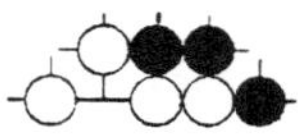 $\}$

$$= \{ \; D \quad , \quad 2* \quad | \quad \{2|0\} \quad , \quad A \; \}$$

Since $2* = \{2 \mid 2\} \geq \{2 \mid 0\} = D$, we can delete the dominated Left option to obtain the simpler form,

$$G = \{2* \mid \{2 \mid 0\}, A\}$$

Neither of the Right options dominate the other, so neither can be deleted. However, when we test for reversibility, we compare G with $G^{RL} = 2$, and we find that $G \leq 2$ because Right, playing second, wins from

$$G - 2 = \{* \mid \{0|\text{-}2\}, A - 2\}$$

This is an easy game; Left must open to $*$ and Right then plays from $*$ to 0 and wins. Thus, Right's move from G to $\{2 \mid 0\}$ reverses through 2 to the Right follower of 2, which is the empty set. Hence,

$$G = \{2* \mid A\}$$

Continuing our tests for reversibility, we compare G with $G^{RL} = A^L = \{3 \mid 1\}$, and we find that $G \leq G^{RL} = \{3 \mid 1\}$, because Right, playing second, can indeed win from

$$G + \{\text{-}1|\text{-}3\} = \{2* \mid A\} + \{\text{-}1|\text{-}3\}$$

If Left opens from $\{2* \mid A\}$ to $2*$, Right plays from $\{\text{-}1|\text{-}3\}$ to -3 and wins because $2* - 3 \leq 0$. If instead Left opens from $\{\text{-}1|\text{-}3\}$ to -1, Right responds by playing from $\{2* \mid A\}$ to $A = \{\{3 \mid 1\} \mid 0\}$ and wins because Left must then play from $A - 1 = \{\{2 \mid 0\} \mid \text{-}1\}$ to $\{2 \mid 0\}$ from which Right plays to 0 and wins. Hence, Right's move from G to A is reversible through $A^L = \{3 \mid 1\}$ to $A^{LR} = 1$, and we have the simplification

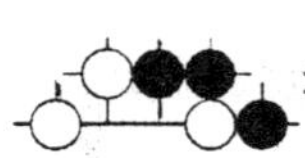 $= \{2* \mid 1\}$

This expression is the desired canonical form. We often find it convenient to omit the braces, and write simply

$$2* \mid 1$$

Integer Translations. It is not hard to verify that

$$3 \mid 1 = 2 + 1 \mid \text{-}1$$

For, if we apply the definition of the sum to the right hand side, we obtain

$$
\begin{aligned}
2 + 1|\text{-}1 & \\
&= G+H \\
&= \{\ G^L + H\ ,\ G + H^L \mid G^R + H\ ,\ G + H^R\ \} \\
&= \{\ 1 + 1|\text{-}1\ ,\ 2 + 1 \ \|\ \ ,\ 2 - 1\ \} \\
&= \{\ 1 + 1|\text{-}1\ , \\
&= 3|1
\end{aligned}
$$

because $3 > 1 + 1|\text{-}1$.

Similarly,

$$
\begin{aligned}
4 \mid 2 &= 3 + 1 \mid \text{-}1 \\
5 \mid 3 &= 4 + 1 \mid \text{-}1
\end{aligned}
$$

etc.

The form shown on the right hand side is more convenient for addition. It is not hard to verify that $1 \mid \text{-}1 + 1 \mid \text{-}1 = 0$. Using this fact, we can compute

$$
\begin{aligned}
& 3|1 + 4|2 + 5|3 \\
= {} & 2 + 1|\text{-}1 + 3 + 1|\text{-}1 + 4 + 1|\text{-}1 \\
= {} & 2 + 3 + 4 + (1|\text{-}1 + 1|\text{-}1) + 1|\text{-}1 \\
= {} & 9 + 1|\text{-}1 \\
= {} & 10|8
\end{aligned}
$$

Each of the above expressions corresponds to a Go endgame. For example, the canonical form of

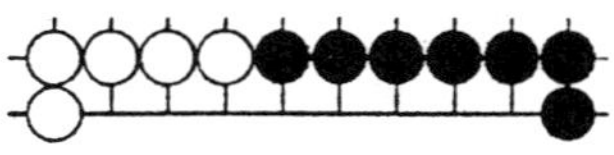

is $3|1$. Similarly, the canonical form of

is $2|0$

Evidently, we also have

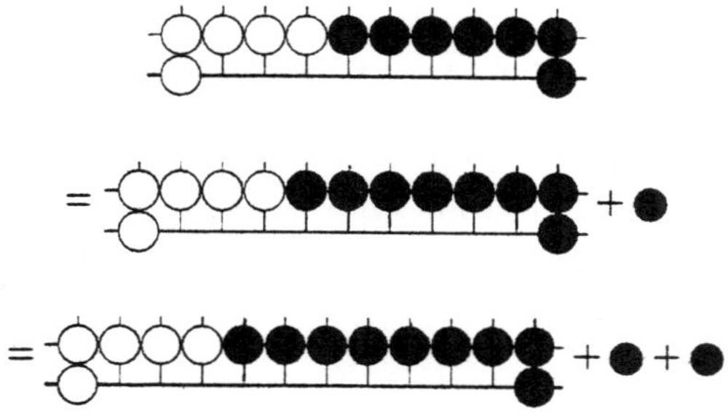

Since we shall frequently encounter similar sums of Go endgames and integers, we find it convenient to define a special notation to deal with them. We shrink the captive stones to a much smaller size called *markings*, and then place them on the board:

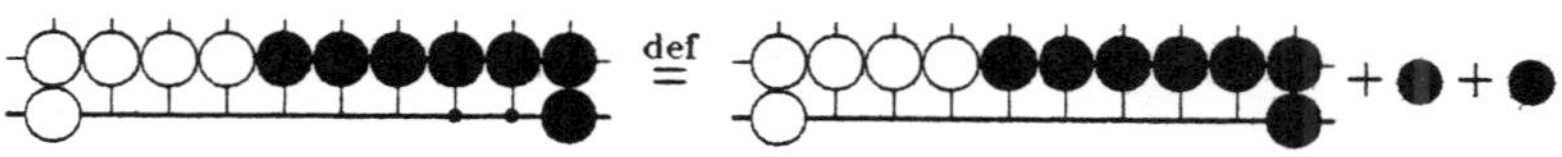

We may also shrink White captive stones:

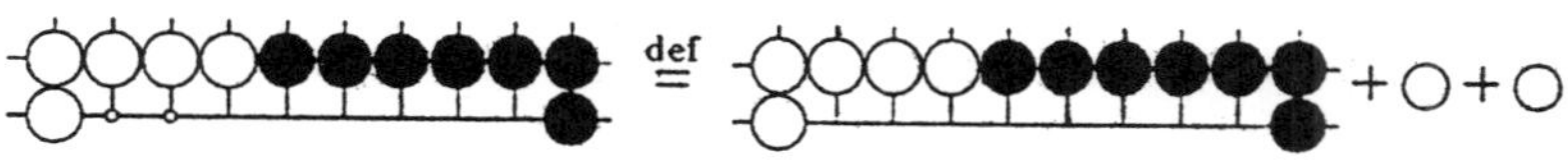

or both:

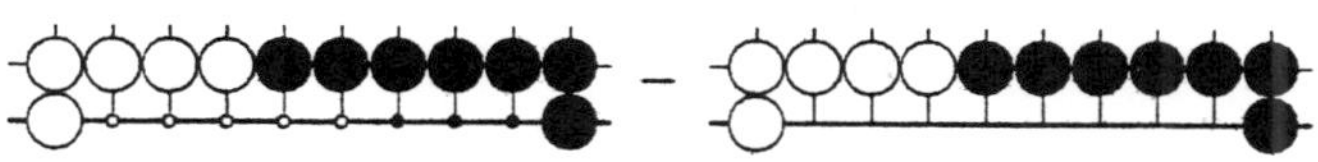

$=$ ○ + ○ + ○ + ○ + ○ + ● + ● + ●

$= 5 - 3 = 2$

We sometimes also show markings atop stones of the opposite color:

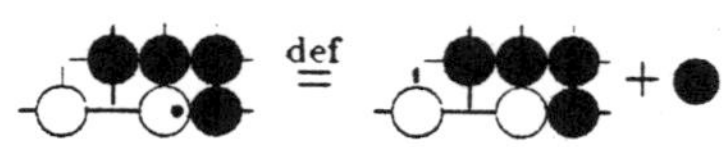

We adhere to the convention of *never* showing shrunken captives atop *immortal* stones.

We may sometimes also show two markings atop a single stone:

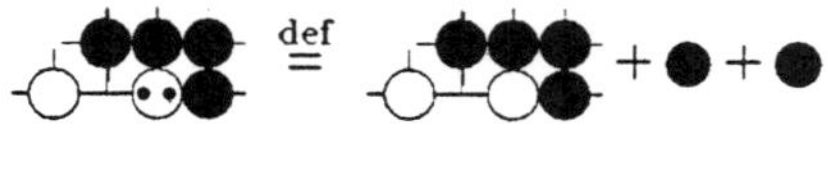

Standard Translations. Since we regard markings merely as a notational convenience, there is no uniquely "correct" way to mark any Go endgame. However, since we are very interested in *sums* of Go endgames, it is usually convenient to separate each endgame into two components - one of which is an integer and the other of which is a residual which is as "small" as feasible, in the sense that even if we add up many copies of the residual, it is still bounded between relatively small integers. For example, we have shown it convenient to represent $3 \mid 1$ as $2 + 1 \mid -1$. The integer component, 2, may be regarded as the "approximate score" of the game $3 \mid 1$. It is certainly correct on the average, for if one takes the sum of an even number of copies of the game $3 \mid 1$, the result is the same as the sum of the same number of copies of the integer 2.

Many good Go players use mental markings to find approximate scores. Empty nodes deep inside a player's territory are marked with his color. Dead stones are marked twice (once for the fact that the stone itself will become a captive, and a second marking for the node beneath it). When a small territory is totally enclosed by stones of the same color, then the standard marking typically yields a game of value 0. One example is nodes D2, E2, F2, G2 and C1, D1, E1, F1, G1, H1 of figure 1:

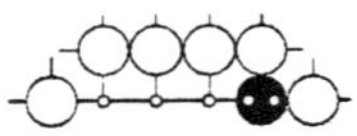 $= 0$

When the boundary of a territory is contested, then the standard marking usually leaves two empty nodes there unmarked. Here are some common Go endgames with markings I regard as "standard."

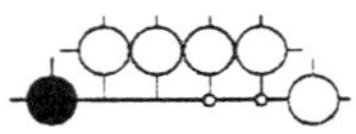

$$= 2* \mid 1 \| 0 ||| -1$$
$$= 2 + \{* \mid -1 \| -2 ||| -3\}$$

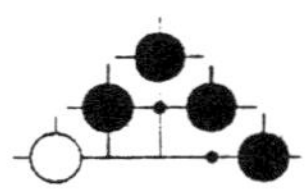

$$= 1 \| 0 \mid -2$$
$$= -2 + \{3 \| 2 \mid 0\}$$

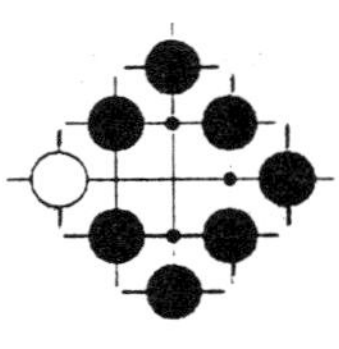

$$= 1 \| 0 \mid -3*$$
$$= -3 + \{4 \| 3 \mid *\}$$

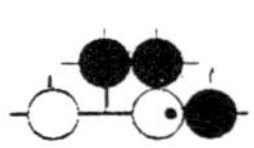

$$= 1 \mid -1$$
$$= -1 + 2 \mid 0$$

The creative Go playing reader will no doubt find an appropriate rationale to justify placing one marking on the white stone. I prefer the mathematical rationale: $1|\text{-}1$ is smaller, on the average, than any other translate of $2|0$. Hence we need one black marking, and it is appropriate to put it as deep into Black's potential territory as possible: atop the White stone rather than atop the empty node.

The Japanese "dame" requires no marking:

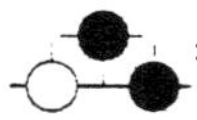 $= *$

A few other (relatively rare) endgames require an odd number of unmarked nodes:

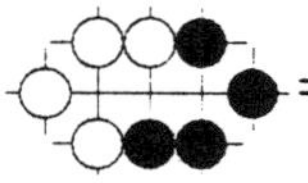 $= 1*|\text{-}1*$

I regard leaving all three nodes unmarked as the "standard marking." One could also achieve the same value by marking opposite ends of the 3 nodes with opposite-colored markings, but this would leave only one empty node, suggesting a less active (i.e. "cooler") value.

Numbers. Numbers are games whose canonical form contains only positions which are numbers, and in which every Left follower is strictly less than every Right follower.

Infinitesimals.
A game G is said to be *infinitesimally close* to a number x iff G exceeds all numbers less than x and G is exceeded by all numbers greater than x. A game which is infinitesimally close to 0 is said to be an infinitesimal.

Numbers and infinitesimals play powerful roles in the solutions of many of the games in Winning Ways. However, it is widely believed that the only numbers which occur directly in Go are integers, and the only infinitesimals which occur directly in Go are 0 and $*$. For that reason, many fundamental abstract games, including ups, downs, and tinies, are virtually unknown among Go players.

A major result of the present work is to show that although the connection is sufficiently subtle to have escaped previous notice, these classical games of Winning Ways hold the key to the analysis and solution of many common nontrivial Go endgames. The correspondence is surprisingly direct, although it entails two mappings, "chilling" and "warming." The former is a special case of the cooling mapping of Winning Ways:

DEFINITION OF COOLING. We recall from Winning Ways that a game G can be mapped into a "cooler" game, G_t, (called "G cooled by t") by the following mapping:

$G_t =$ $t + G^L{}_t \mid -t + G^R{}_t$ unless there is some τ, $0 \leq \tau \leq t$, such that G_τ is infinitesimally close to a number, x, in which case $G_t = x$

THEOREM:

$$(G+H)_t = G_t + H_t$$

If all of the positions of a game have negatives, then if that game is cooled by a sufficiently large amount (called "freezing") then the game becomes a number (called the "mean value"). Many loopy games can also be frozen. For example, the reader may verify that

$$\textcircled{K}_{1/3} = 1/3 + \textcircled{k}$$

where $\textcircled{k}$ and its complement satisfy

$$\begin{aligned} \textcircled{k} &= \{\overline{\textcircled{k}} \mid 0\}; \\ \overline{\textcircled{k}} &= \{0 \mid \textcircled{k}\} \end{aligned}$$

$\textcircled{k}$ and its complement belong to a class of infinitesimals called "all-smalls." Since they are infinitesimals, it follows that $\textcircled{K}$ has freezing temperature of 1/3 and mean value [5] of 1/3.

Since cooling is a many-to-one mapping, its inverse mapping is not unique. Among the many mappings which attempt to invert cooling, I consider the following to have been the most successful. It defines $\int_s^t G$ (read "G overheated from s to t") as follows.

DEFINITION OF OVERHEATING:

$$\int_s^t G = \begin{cases} \overbrace{s+s+s+\cdots+s}^{G \text{ times}} & \text{if } G \text{ is a nonnegative integer} \\ -\int_s^t (-G) & \text{if } G \text{ is a negative integer} \\ t + \int_s^t G^L \big| -t + \int_s^t G^R & \text{otherwise} \end{cases}$$

In "Blockbusting and Domineering," variations of this definition were used to express the values of an infinite number of $2 \times n$ and $3 \times n$ Domineering positions in terms of appropriately overheated numbers.

In Go endgames, the most useful cases of cooling and overheating are the following:

DEFINITION OF CHILLING: Chilling is Cooling by one degree ($t = 1$).

DEFINITION OF WARMING: Warming is overheating from $1*$ to 1.

---

5 Although Winning Ways' main theorem about mean values is not valid for games like $\textcircled{K}$ when played according to the rules of Basic Mathematical Go, there are more sophisticated dialects of mathematical rules under which the validity of a modified version of this theorem is restored. In fact, under an appropriate set of rules, the fact that $\textcircled{K}$ has mean value 1/3 is very convincingly demonstrated by the equation $\textcircled{K} + \textcircled{K} + \textcircled{K} = 1$.

THEOREM: *When applied to games whose canonical forms are loop-free, warming is both linear and monotonic.*

One elegant proof is simply to observe that warming coincides with Norton multiplication by the game 1∗. [c.f. Winning Ways, pages 246-248].

BIG THEOREM: *For "normal" Go endgames (whose canonical forms are loop-free), chilling can be inverted by warming and then adding 0 or ∗, according to an appropriate parity.*

Because chilling, in general, is many-to-one, its invertibility may come as something of a mathematical surprise. However, the fact that all stopping positions of normal Go endgames are sums of integers and stars provides sufficient constraints to resolve the ambiguity.

Since chilling is linear, monotonic, and invertible, there is a direct correspondence between normal Go endgames and their chilled images. The values of the chilled images are often old friends from Winning Ways, such as numbers, stars, ups, downs, and tinies.

We chill so many Go endgames so often that we find it convenient to introduce a special notation to denote the chilled game: we run a boundary line through the centers of all of the (necessarily immortal) stones on the edge of the region, and delete everything outside of the boundary. Here are some examples:

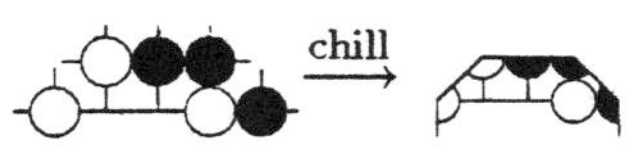

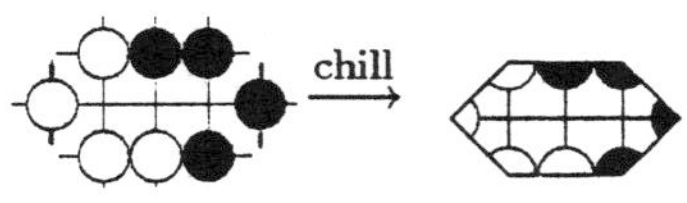

The chilled game may have markings just like the game from which it came:

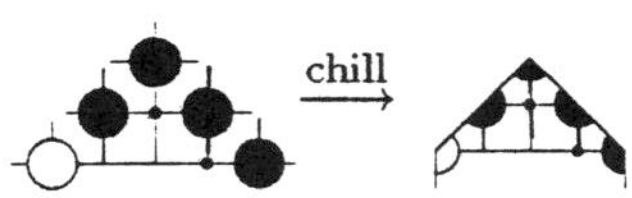

With practice, one can play "chilled Go" directly. When the value of the game to be chilled is not a number, it is played just like Go, except that there is a one-point "tax" on every move. Those who play "paper and pencil chilled Go" usually find it convenient to settle their tax accounts by appropriate adjustments of the markings on the board. Thus,

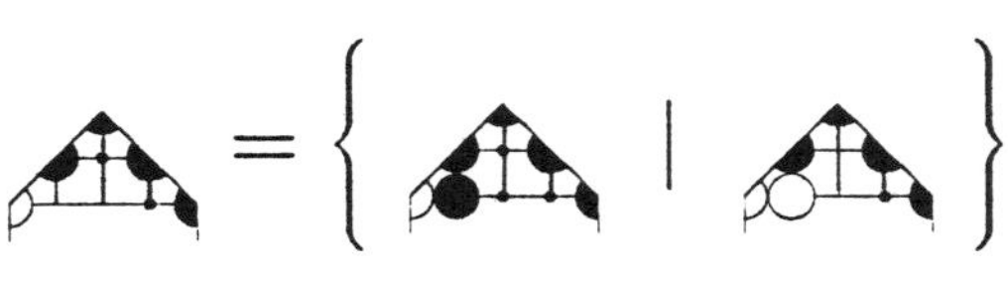

and

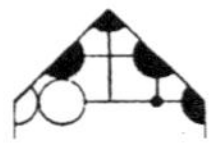 = {  |  }

 $= 0$

whence

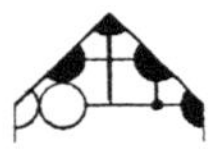 $= *$

and

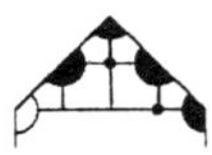 $= 0|* = \uparrow$

Here are six more chilled Go endgames and their values:

 $= 0 \mid -\frac{1}{2}$

 $= \Uparrow *$

 $= *$

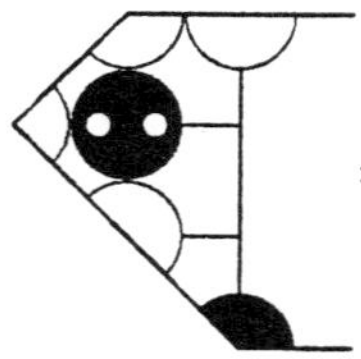

$= \downarrow$

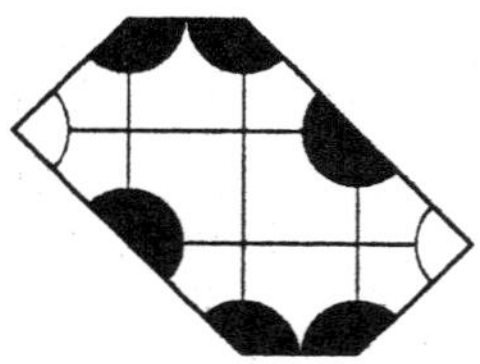

$= \frac{1}{2}$

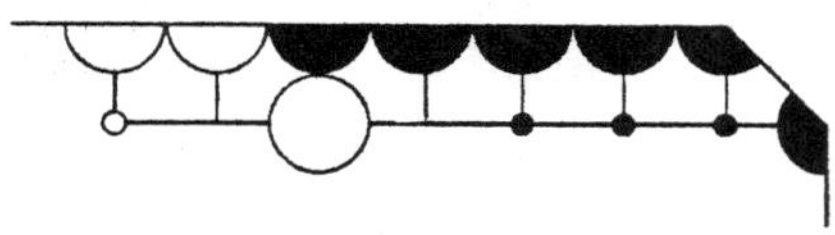

$= \downarrow$

The value of the sum of these six games is $\frac{1}{2} \mid 0$, which can be won by the first player. This sum is the "solution" to the problem we presented earlier, as shown in figure 2 in which the lines dividing the board into regions are assumed to immortalize all of the stones they cut.

In order to make things hard for White, Black should always play a dominant move, as in figure 3. Starting from figure 1, the next 17 moves are played the same in all modern versions of the rules of Go.

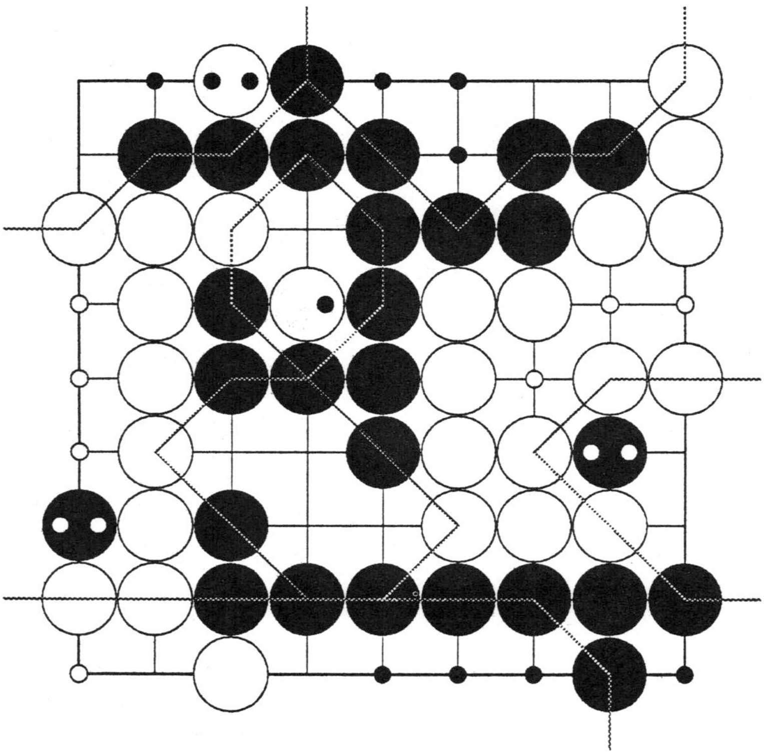

FIG. 2. *Mathematical solution*

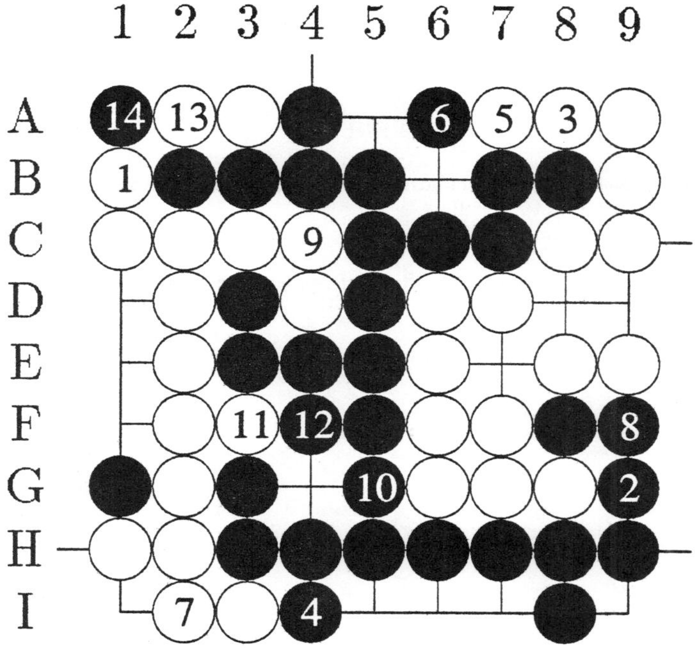

FIG. 3. *A well played endgame*

Black's 14 captures the White stones at A2 and A3: White then plays 15 at A2, capturing the Black stone at A1. Then 16 at A3; 17 at A1. The game has then reached a "stopping position," from which all moves of both players have mathematical incentive of -1.

There are two white and one black captured stones off the board and a dead Black stone at G1. Black has 7 points of territory: A5, B6, G4, I5, I6, I7, I9. White has 8 points of territory: D1, E1, F1, G1, D8, D9, E7, I1.

To bring the mathematical game to an end: White, on successive turns, should capture dead Black stones, then fill in all of his own territory, and finally return captured Black stones to the pot. This will ensure that Black must run out of legal moves before White does, so White will win. White has *only one* winning move at each of 1, 3, 5, 9, 11, 13, 15, and 17, and several of these are rather hard to find without mathematical methods.

REFERENCES

[1] E. R. Berlekamp, "Blockbusting and Domineering", *J. of Combin. Theory, Ser. A* **49**, 67-116 (1988).

[2] E. R. Berlekamp, J. H. Conway, R. K. Guy, *Winning Ways*, Academic Press, (1982).

[3] J. Davies, R. Bozulich, *The Rules of Go*, Ishi Press International, Mountain View, CA, (1986).

Proceedings of Symposia in Applied Mathematics
Volume **43**, 1991

Games and Codes

Vera Pless

1. Introduction. Combinatorial games have an important connection with numbers [1, 3]. They can also be expressed nicely as graphs [8]. It was noticed [1, p.434, 8; p.40] that winning positions in certain coin-turning games are really vectors in particular well-known codes. The connection of games with error-correcting codes was greatly strengthened by the recent Conway-Sloane construction of lexicodes [6], error-correcting codes whose codewords are winning moves in a game. Our aim is to describe many of these connections and to reveal the similarity of important coding concepts with important game concepts. As many of the papers in this series describe combinatorial games our description of games will be brief. We will define codes and also give properties of codes which are interesting for the codes themselves and for their associated games. These properties concern the group of a code and combinatorial designs held by vectors of a fixed weight in a code. Our discussion will raise several open problems.

2. Codes. As binary codes have a special role in the codes related to games we will start with these codes. More details and further discussions can be found in [9].

An (n, k) *binary code* is a k-dimensional subspace of the space of binary n-tuples. Hence it can be described by a basis which in coding terms is called a *generator matrix.*

An important concept in coding is the *weight* of a vector, that is the number of non-zero components it has. The weight of x is denoted by $wt(x)$. The smallest non-zero weight of a vector in a code is called the *minimum weight* of the code and is often denoted by d. When the minimum weight d is known we say we have an (n, k, d) code. For example the following is a generator matrix for a $(7, 4, 3)$ code called the Hamming code C.

$$G = \begin{pmatrix} 0 & 1 & 1 & 0 & 1 & 0 & 0 \\ 0 & 0 & 1 & 1 & 0 & 1 & 0 \\ 0 & 0 & 0 & 1 & 1 & 0 & 1 \\ 1 & 0 & 0 & 0 & 1 & 1 & 0 \end{pmatrix}.$$

Supported in part by AFOSR Grant 880237.

This paper is in final form and no version of it will be submitted for publication elsewhere.
1980 *Mathematics Subject Clasification* (1985 *Revision*). Primary 90D05; Secondary 94B25.

The *weight distribution* of a code is the number of vectors of any weight in the code. The number of vectors of weight i in a code is denoted by A_i. As any code must contain the all zero vector we always have $A_0 = 1$. It is easy to see that the weight distribution of the Hamming code C is $A_3 = A_4 = 7$, $A_0 = A_7 = 1$ and the other A_i are zero. Note that we have $7 + 7 + 1 + 1 = 2^4$ vectors in C.

Two codes are called *equivalent* if one can be obtained from the other by a permutation of the coordinate indices. It is not hard to see [9, p.166] that any two $(7, 4, 3)$ binary codes are equivalent.

If C is an (n, k) code, we define the *extended code* of C, $\widehat{C}$ to be the $(n+1, k)$ code with an overall parity check added, this is a 0 if the codeword has even weight and a 1 if it has odd weight. Clearly all vectors in $\widehat{C}$ have even weight. A generator matrix of the extended Hamming code $\widehat{C}$ is the following.

$$\widehat{G} = \begin{pmatrix} 1 & 0 & 1 & 1 & 0 & 1 & 0 & 0 \\ 1 & 0 & 0 & 1 & 1 & 0 & 1 & 0 \\ 1 & 0 & 0 & 0 & 1 & 1 & 0 & 1 \\ 1 & 1 & 0 & 0 & 0 & 1 & 1 & 0 \end{pmatrix}.$$

If C has odd minimum weight d, then $\widehat{C}$ has minimum weight $d+1$. Again it is not hard to see that there is a unique (up to coordinate permutation) $(8, 4, 4)$ code. The weight distribution of $\widehat{C}$ follows from that of C and it is $A_0 = A_8 = 1$, $A_4 = 14$.

Interesting codes have been associated to interesting groups and interesting combinatorial designs. The *group of a binary code* is the set of all coordinate permutations which send codewords onto codewords. The Hamming code C described above with coordinates labelled $0, 1, \dots, 6$ is a *cyclic code*; this means that the coordinate permutation $i \to i+1 \pmod 7$ sends a codeword into another codeword. The coordinates of a cyclic code of length n are always labelled $0, 1, \dots, n-1$. As $A_3 = 7$ all vectors of weight 3 in C are cyclic shifts of one weight 3 vector. Similarly for the weight 4 vectors since $A_4 = 7$. Thus we can easily recognize all the vectors in C as the only other vectors are the all zero vector and the all one vector. The vectors in $\widehat{C}$ can be easily recognized from recognizing vectors in C. As C is a quadratic residue code, the coordinate permutation $i \to ai \pmod 7$ where a is a fixed square in $GF(7)$ also leaves C invariant. The group of $\widehat{C}$ (with the extended coordinate labelled ∞) also contains the coordinate permutation $i \to -1/i \pmod 7$ which is understood to exchange 0 and ∞.

We can play a game with a row of 7 coins. If we consider tails as zeros and heads as ones, this is a binary 7-tuple. Each player turns one or two coins (with certain specifications to make sure the game ends) and the one who cannot make a move loses. The specification that we always move to a lexicographically smaller 7-tuple ensures that the last position is the all zero vector. If a winning strategy were to always play into vectors in the code, then the ability to recognize all the code vectors would be very useful. This does happen, but we would like to make a few more remarks about the similarity of such moves and important coding concepts.

The *distance*, $d(x, y)$, between two binary vectors x and y is the number of positions in which they differ; that is $d(x, y) = wt(x + y)$. The *covering radius*

r of a code is the furthest distance from a vector outside the code to a closest codeword. In other words $r = \max_x\{\min d(x,y) \text{ for } y \text{ in } C\}$. The covering radius is an important parameter of a code and is known for some codes. The covering radius of the Hamming $(7,4,3)$ code C is one. This means that if x is a vector outside C we can reach a vector in C by changing one coordinate of x which is the same as adding a weight one vector to x. If we regard x as a row of 7 coins, this means that if x is not in C, then we can reach a vector in C by turning one coin. Notice that in our game we could turn either one or two coins.

Error-correcting codes were originally devised to correct errors in the transmission of digitally encoded information. We suppose that a message denoted by c from the code is sent and some errors occur in transmission. We represent this by an error vector e which has ones at the positions in error and zeros elsewhere. Then the received message $x = c + e$ is not in the code. (Sometimes many errors could result in another codeword but we would not be able to detect such errors.) Then *decoding* x is finding the closest vector in the code to x. This is the same as adding an error vector e of smallest weight to x so that $e + x$ is in the code. So when we turn coins to reach a code vector this process is similar to decoding. However, we are not always looking for an error vector of smallest weight as we will see later. We want to be sure that our game will end.

3. Lexicodes. Lexicodes were defined by Conway and Sloane [6]. The coin-turning games known to be associated to codes [1, 8] were shown to be lexicodes with certain parameters. Even more surprising binary lexicodes are binary codes, that is linear spaces over $GF(2)$. This is really unexpected as lexicodes are defined in terms of heap games.

If we label the coordinate positions of our seven coins with the numbers 1 through 7, we can regard these as heaps of sizes 1 through 7 in nim with a heap being present if the coin has a head in that position and not present if it has a tail. This configuration can also be represented by a binary 7-tuple. Then [1, p.430] the authors show that turning one or two coins is the same as playing nim with heap sizes up to 7. Please note that this labelling of the coordinate positions is different from the labelling of the positions of the Hamming code considered as a quadratic residue code.

According to [6] a heap game has certain atomic positions P_i and general position

$$P_a + P_b + P_c + \dots$$

The rules of the game are given by a family of turning sets and a legal move is to replace any term P_h by $P_i + P_j + \dots$ provided that $\{h, i, j, \dots\}$ is a turning set. By the well-known theory of Sprague and Grundy there is a function $G(P)$ (the nim-value or Grundy number) assigning integer values to positions P with the property that a player wins by consistently moving to positions of G-value zero.

Conway and Sloane go on to note that the general position $\sum_{i=0} n_i P_i$ in such a game may be represented by the integral vector

$$(\dots n_3 n_2 n_1)$$

$$\text{(or } (\dots n_3 n_2 n_1 n_1) \text{ if zero heaps play a role).}$$

Hence the winning strategy can be represented by a binary code consisting of the binary vectors

$$(\dots \zeta_3 \zeta_2 \zeta_1),$$

for which the nim-sum

$$(*) \qquad \sum \zeta_i G(P_i) = 0.$$

These vectors clearly form a binary linear code which the authors call the winning code. In view of (*) Conway and Sloane [6] regard a heap game as played with binary numbers $N = \sum \zeta_i 2^i$ (or binary vectors $(\dots \zeta_3 \zeta_2 \zeta_1)$) where a legal move is to replace N by $N' = \sum \zeta_i' 2^i$ provided

1) $N' < N$.

This is called the lexicographic condition and ensures that the game will end, and

2) the set of i is so that $\zeta_i' \neq \zeta_i$ is a turning set.

For each family of turning sets and each base B Conway and Sloane then define lexicodes by a greedy algorithm. We just give their definition for $B = 2$.

We construct all binary words $(\dots \zeta_3 \zeta_2 \zeta_1)$ in the lexicographic order determined by the corresponding number

$$N = \sum \zeta_i B^i.$$

As it is constructed each word is considered for inclusion in the lexicode. A word is rejected if there is some earlier word $N' = (\dots \zeta_3' \zeta_2' \zeta_1')$ for which the set of i with $\zeta_i' \neq \zeta_i$ is a turning set, and is accepted otherwise, that is placed in the code. It turns out that this code is the same as the winning code. The reader should consult [6] for further details.

Notice that if we take the turning sets to be all sets of size t or smaller, then we are constructing a code with minimum distance $d = t + 1$. As an example we give the winning code, the lexicode, of length 7 constructed in this greedy fashion with the turning sets all sets of size 1 or 2.

	7	6	5	4	3	2	1
1	0	0	0	0	0	0	0
2	0	0	0	0	1	1	1
3	0	0	1	1	0	0	1
4	0	0	1	1	1	1	0
5	0	1	0	1	0	1	0
6	0	1	0	1	1	0	1
7	0	1	1	0	0	1	1
8	0	1	1	0	1	0	0
9	1	0	0	1	0	1	1
10	1	0	0	1	1	0	0
11	1	0	1	0	0	1	0
12	1	0	1	0	1	0	1
13	1	1	0	0	0	0	1
14	1	1	0	0	1	1	0
15	1	1	1	1	0	0	0
16	1	1	1	1	1	1	1

This is precisely the winning code for turning turtles [6]. The column labels indicate the heap sizes for nim. These heaps are considered present whenever the vector has a one. Thus the second vector represents three heaps of sizes 1, 2 and 3. Notice that their nim sum is zero. This is true for any vector in this winning code. Notice also that if x is any vector outside of our code we can reach a codeword by adding a vector of weight one and a codeword of lexicographically smaller value by adding either a vector of weight one or a vector of weight two. We can see that this must be so, otherwise we would have chosen the vector x to be in the code of this coin-turning game. This observation holds in the general case and is the basis of the proof that the lexicode is the winning code of its game.

As we noted binary lexicodes are actually binary linear codes. They seem to be quite good ones from the data in table VII [6] for binary (n, k, d) lexicodes with $n \leqq 44$ and $d \leqq 10$. That is, the codes in this table either have the parameters for the best (n, k, d) codes known or come within one or two dimensions of the best ones. Given n and d we can construct a binary lexicode but little seems to be known theoretically about k. The values of k in table VII were, I imagine, computed by constructing the entire code. Can anything be said about k a priori? It is also interesting that lexicodes over $GF(2^{2^a})$ are linear codes [6]. Lexicodes over other bases B are not linear codes except for occasional accidental exceptions. Conway [4] turns this situation upside down and assumes that lexicodes constructed in a certain greedy way are linear. This leads to very interesting results about what their fields can be.

4. Certain Lexicodes and Quadratic Residue Codes. In [6] some well-known codes are identified as lexicodes. These codes merit further discussion. All the following codes are binary.

1) Zero-sum codes: Here n is anything and $d = 2$. This is the $(n, n-1, 2)$ code of all even-weight vectors.

2) When $d = 3$ and $n = 2^m - 1$, the authors [6] state that the lexicodes are Hamming codes. The Hamming codes are $(2^m - 1, 2^m - 1 - m, 3)$ codes. When $n = 7$, this is the Hamming code described above and the corresponding game is turning turtles [1, p.429]. All of these codes are known to have covering radius one. As we noted before if we start with any vector outside the code we can reach a codeword by turning one coin which is the same as adding a vector of weight one. The authors' results show, for all of these codes, that we can reach a lexicographically smaller codeword by adding a vecto. of weight either one or two.

When $d = 3$ for other values of n, the lexicodes are identified as shortened Hamming codes.

3) Extended codes:

If we want to construct a lexicode with $n = 8$ and $d = 4$ we could do it easily using the example we already constructed for $n = 7$ and $d = 3$. We would just have to add a new position to the right and an overall parity check. This is true in general so that lexicodes with d even are obtained from lexicodes with d odd by adding an overall parity check. This observation is the same as the Mock Turtle Theorem [1, p.432].

When $n = 8$ and $d = 4$ the heap game is called Mock Turtles and the new parity position is labelled 0 when we interpret the coordinate labels to be heap sizes.

4) The extended quadratic residue code of length 18.

Consider the game with 17 coins which allows turning 1, 2, 3, or 4 coins with legal moves to lexicographically smaller vectors. The winning code for this game is equivalent to a $(17, 9, 5)$ code called the quadratic residue code. If we use 18 coins and allow up to 5 turns with the lexicographic condition, by the discussion above the winning code is the $(18, 9, 6)$ extended quadratic residue code. The corresponding game is called Moebius [1, p.435].

5) The extended Golay code.

When $d = 7$ and $n = 23$ the lexicode is the $(23, 12, 7)$ Golay code which is also a quadratic residue code. For $d = 8$ and $n = 24$ we get the extended $(24, 12, 8)$ Golay code. The corresponding game is called Mogul [1, p.435].

6) Binary codes with distance 10.

The dimensions of lexicodes are known for $n = 27$ or 31 and these codes have parameters $(27, 9, 10)$ and $(31, 12, 10)$ respectively. The corresponding game is Moidores.

The natural question here is are there other interesting lexicodes? As several of these codes are quadratic residue codes or extended quadratic residue codes we will say something about these codes now.

Binary quadratic residue codes [9, p.94] exist of length p where p is a prime $\equiv \pm 1 \pmod 8$. We label the coordinate indices from 0 to $p - 1$. Quadratic residue codes are all cyclic codes so that whenever a vector is in the code so are all its cyclic shifts just as for the Hamming $(7,4,3)$ code. Note that the $(7,4,3)$ lexicode whose codewords we listed is not in the quadratic residue representation but must be equivalent to the quadratic residue $(7,4,3)$ code as a $(7,4,3)$ code is unique up to coordinate permutation.

The fact that a quadratic residue code is cyclic is the same as saying that the coordinate permutation $i \to i+1 \pmod p$ leaves the code invariant. Other permutations are known to be in the group of a quadratic residue code and we describe them. All coordinate permutations of the form $i \to ai \pmod p$ where a is a fixed square in $GF(p)$ leave each quadratic residue code of length p invariant. There are exactly four quadratic residue codes of length p for p a prime $\equiv \pm 1 \pmod 8$. Two of these are equivalent $(p, \frac{p+1}{2})$ codes and two are equivalent $(p, \frac{p-1}{2})$ codes [9, sec. 6.4].

We label the additional coordinate position of the extended quadratic residue code by ∞. Then the permutations described above on the coordinates $0, \dots, p-1$ which fix ∞ are in the group of the extended quadratic residue code. The coordinate permutation $i \to -1/i \pmod p$ (which interchanges 0 and ∞) is also in the group of the extended quadratic residue code. The permutations described generate the group $PSL_2(p)$ of order $\frac{(p-1)p(p+1)}{2}$ [9, p.98]. The group of the extended quadratic residue $(8,4,4)$ is larger having order 1,344. The group of the extended $(24,12,8)$ Golay code is much larger being the Mathieu group M_{24}.

Codewords in a quadratic residue code are easy to locate as a generator matrix can be constructed from one vector whose ones are determined by the squares in $GF(p)$ and its cyclic shifts [9, pp. 95 and 97]. The weight distributions of the $(8,4,4)$, $(18,9,6)$ and $(24,12,8)$ extended quadratic residue codes are known. The weight distribution of the $(8,4,4)$ code has already been given. The weight distributions of both the $(18,9,6)$ and $(24,12,8)$ codes are in [1, p.435]. Also in this reference is a description of all vectors in the $(18,9,6)$ code represented as an extended quadratic residue code. It is possible to do this with the 2^9 codewords as one only has to describe one vector for each cycle class of 17 vectors. A modified form of the MOG by R. Curtis is also in [1]. The MOG is a device which enables players to find the unique weight 8 codeword in the extended Golay code which contains any 5 given points.

As noted before finding a closest vector in the code is the decoding problem. A hand-decoding scheme for the extended $(24,12,8)$ Golay code is given in [7, chap. 11, 10]. The covering radius of the extended Golay code is 4 and this decoding scheme easily locates a codeword at distance 1, 2, 3 or 4 from any given vector outside the code. It can also be used to identify codewords. However, a word of caution is in order. In order to play Mogul we must not only locate a codeword but a codeword which is lexicographically smaller than the vector we started with. Even though we can turn as many as 7 coins not just 4, it is not clear to me how we might do this. I am worried about the analogous problem for Moebius. We can even see this problem by looking at the two versions of the $(7,4,3)$ Hamming code.

It is interesting to note that the hand-decoding scheme mentioned above for the extended Golay code is based on a construction of this code from the $(6,3,4)$

hexacode over $GF(4)$. This latter code is also a lexicode [6].

Naturally one wants to know if there are other lexicodes which are well-known codes. Such knowledge could give information about the group of the lexicode. There are extended quadratic residue $(32, 16, 8)$ and $(42, 21, 10)$. According to Table VII [6] there are $(32, 16, 8)$ and $(42, 20, 10)$ lexicodes. The latter cannot be an extended quadratic residue code. Is the $(32, 16, 8)$ lexicode an extended quadratic residue code? There are at least four other known inequivalent codes with these parameters [5]. It can be shown that the $(32, 16, 8)$ lexicode cannot be any of these [11]. The codes mentioned in [5] including the $(32, 16, 8)$ extended quadratic residue code are all self-orthogonal codes. But by the greedy lexicographic algorithm the first 12 dimensions of the $(32, 16, 8)$ lexicode will be the extended $(24, 12, 8)$ Golay code in its lexicode form with 8 zeroes appended to the beginning of each vector. The last 4 dimensions of the $(32, 16, 8)$ lexicode cannot be orthogonal to the first 12 dimensions as no more vectors can be orthogonal to the Golay code. Hence this lexicode is not self-orthogonal and so cannot be equivalent to a self-orthogonal code.

Quadratic residue codes have been generalized to m-adic residue codes [2]. Are any of these lexicodes? As Table VII ends with $n = 44$, we cannot determine whether there are longer lexicodes which could be quadratic residue codes.

5. Designs, Constant Weight Codes and Welters Games. A $t-(v, k, \lambda)$ design is a set of v points and blocks each consisting of k points with the property that every t points are contained in exactly λ blocks of the design. We can regard a block as a binary vector with ones at the points it contains and zeros elsewhere. Clearly a block has weight k. It sometimes happens that the vectors of a fixed weight in a code of length v constitute the blocks of a $t-(v, k, \lambda)$ design. It is very interesting when this happens, particularly when the fixed weight is the minimum weight. Designs with $\lambda = 1$ are especially interesting so that a $t-(v, k, 1)$ design is called a Steiner system and denoted by $S(t, k, v)$. The vectors of weight 3 in the Hamming $(7, 4, 3)$ code form an $S(2, 3, 7)$. The vectors of weight 4 in the extended Hamming $(8, 4, 4)$ code are blocks of an $S(3, 4, 8)$. The vectors of any fixed weight in the extended $(18, 9, 6)$ quadratic residue code are the blocks of a 2-design. Of course the λ's for the different designs are different but they can all be computed from the weight distribution of the code [9, chap. 9].

The vectors of weight 8 in the extended Golay code form the blocks of the famous Steiner system $S(5, 8, 24)$. This explains why given any 5 points there is a unique vector of weight 8 in the Golay code which contains these 5 points. The vectors of the other weights, 12 and 16, in the Golay code also constitute blocks of 5-designs. As the Mathieu group M_{24} is the group of the code it sends vectors of a fixed weight onto themselves and so is the group of the Steiner system $S(5, 8, 24)$ and also the group of the other 5-designs mentioned.

Certain non-linear codes are of interest. *Constant weight* codes are one example. We again consider only binary codes. A constant weight code consists of vectors of a fixed weight w so that the distance between any two of these vectors is at least d. Since the distance between two binary vectors of fixed weight is even, we take d to be even. The set of vectors of any given weight w in a code is an example of a constant weight code where d is either the minimum weight of the code if that is even or one more than the minimum weight of the code if odd. The vectors of a fixed weight in the codes described above are interesting constant weight codes.

The definition of *constant weight lexicodes* is similar to the definition of lexicodes [6]. Consider all vectors of a specified weight w in lexicographic order and accept a word if it has distance at least d from all previously accepted words.

The game corresponding to a constant weight lexicode consisting of vectors of weight w with $d = 2t \geqq 4$ is not described by turning sets but is as follows. A typical position is a set $\{a_1, a_2, \ldots, a_w\}$ of distinct non-negative integers and the legal move is to decrease $1, 2, \ldots$ or $t-1$ of these integers while preserving their distinctness. As for the lexicodes and their games, the winning code for this game is the constant weight lexicode with the same parameters.

When $d = 2$, the corresponding constant weight lexicode consists of all codewords of weight w. The constant weight lexicode for $d = 4$ and any w is the winning code for a game called Welter's game [1, pp. 472-481], [3, ch. 13]. In Welter's game a typical position is a set $\{a_1, a_2, \ldots, a_w\}$ of distinct non-negative integers, and the legal move is to decrease one of these integers while preserving their distinctness. The solution is surprisingly complicated and involves Welter's function which contains both nim sums and ordinary subtraction. Various schemes to calculate the values of this function are given including frieze patterns.

For $d = 4$, $w = 3$ and $n = 2^m - 1$ the constant weight lexicode consists of the vectors of weight 3 in the Hamming code of length $2^m - 1$. If $d = w = 4$ and $n = 2^m$, the constant weight lexicode consists of the weight 4 vectors in the extended Hamming code of length 2^m. This is because Welter's game for $w = 3$ or 4 is equivalent to nim [1, p.473].

When $w = d = 8$ and $n = 24$ the constant weight lexicode consists of the 759 weight 8 codewords in the extended $(24, 12, 8)$ Golay code and so are blocks of a Steiner system $S(5, 8, 24)$ [6]. No other relations with known codes or designs are given for constant weight lexicodes in [6] except when a sum constraint is added and in that situation the vectors of the constant weight lexicode are blocks of a Steiner system $S(5, 6, 12)$.

References

1. E. R. Berlekamp, J. H. Conway and R. K. Guy, *Winning Ways for your Mathematical Plays*, Academic Press, 1982.

2. R. A. Brualdi and V. Pless, *Polyadic codes*, Discrete Applied Mathematics, **25** (1989), 3-17.

3. J. H. Conway, *On Numbers and Games*, Academic Press, 1976.

4. J. H. Conway, *Integral lexicographic codes*, Discrete Mathematics, to appear.

5. J. H. Conway and V. Pless, *On the enumeration of self-dual codes*, JCT **A28** (1980), 26-53.

6. J. H. Conway and N. J. A. Sloane, *Lexicographic codes: error correcting codes from game theory*, IEEE Trans Info. Theory, IT-**32** (1986), 337-348.

7. J. H. Conway and N. J. A. Sloane, *Sphere Packings, Lattices, and Groups*, Springer-Verlag, 1988.

8. R. K. Guy, *Combinatorial Games*.

9. V. Pless, *Introduction to the Theory of Error-Correcting Codes*, second edition, John Wiley and Sons, 1989.

10. V. Pless, *Decoding the Golay codes*, IEEE Trans. Info. Theory, **32** (1986), 561-567.
11. J. Conway, *Private Communication.*

Department of Mathematics, Statistics, and Computer Science, University of Illinois at Chicago, Chicago, Illinois 60680

Proceedings of Symposia in Applied Mathematics
Volume **43**, 1991

Complexity of Games

Aviezri S. Fraenkel

Department of Applied Mathematics and Computer Science
The Weizmann Institute of Science
Rehovot 76100, Israel

This paper has been reprinted with permission from: *From Nim to Go*, in the Annals of Discrete Mathematics, vol. 6 (1980), pp. 137–156.

1980 *Mathematics Subject Classification* (1985 *Revision*). Primary 90D05; Secondary 68C25.

Abstract. Nim, checkers, Go and chess belong to the same family of 2-player full information 0–1 games without chance moves. Yet the strategy of Nim is very simple, that of chess seemingly very hard. Why?? Some of the reasons are explored and new theories are demonstrated by means of sample games of increasing complexity which span a bridge from Nim to chess.

1. Nim-Like Games

Nim is of course a well-known game. Fig. 1 is an example with 3 piles: a pile with 1 token, a pile with 2 and a pile with 3 tokens. Two players alternately select one pile and remove from it any positive number of tokens, possibly the entire pile. The player first unable to move loses and his opponent wins. The theory of the game is very simple, though perhaps unexpected: Write down the number of tokens in each pile in binary, and sum without carry (vector addition over $GF(2)$; also known as XOR addition). If the resulting *Nim-sum* σ is 0, then player II can win (irrespective of the moves of his opponent); otherwise player I can win (irrespective of the moves of player II). For the above example $\sigma = 0$ (Table 1). So it pays to be polite: offer your opponent to make the first move!

Figure 1. A game of Nim with three piles.

Table 1. The Nim-sum of the game in Fig. 1 is 0.

1	1
1	0
0	1
$\sigma = 0$	0

1.1. The Class of Combinatorial Games

Nim is a member of the following family of games: 2-player, perfect-information (unlike some card games where information is hidden) without chance moves (no dice), and outcome restricted to lose, win or draw. The players move alternately. This family — including some scattered attempts at breaking out in several directions, see e.g. Reif (1984), Li (1977) — is loosely known under the name *combinatorial games*, for short *games* below. We shall further assume throughout that play is *normal*, i.e., the player first unable to move loses and his opponent wins. Only at the end of §6.2 we discuss briefly the implications of *misère* play, where the player first unable to move wins and his opponent loses.

The class of combinatorial games is a very respectable family. No game played in Las Vegas or Atlantic City taints the family! Furthermore, well-known games such as checkers, Go and chess are proud family members. Though both Nim and chess belong to the family, Nim has a very simple strategy, but no chess world-champion yet *knows* whether White has an opening move which guarantees a win; a fortiori nobody seems to know whether White can win from an arbitrary mid-game position, which is typically considerably less symmetric than the initial position.

What are the reasons for this discrepancy? This question motivates our investigation of combinatorial games.

1.2. Chess is Just as Easy as Nim

We begin with some notation for games without draws.

Let N be the set of all positions from which the *N*ext (first) player can win, and P the set of all positions from which the *P*revious player can win. With each game we associate its *game-graph* $G = (V, E)$, which is a finite acyclic directed graph (digraph), where V is the set of game positions and $(u, v) \in E$ if and only if there is a move from u to v. See Figs. 2 and 3.

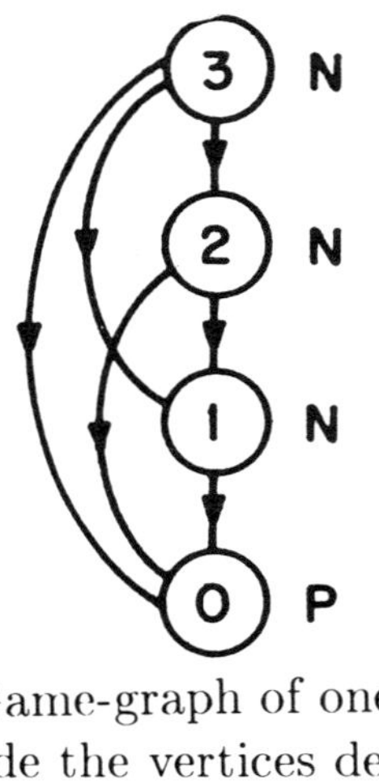

Figure 2. Game-graph of one Nim-pile.
The numbers inside the vertices denote the pile size.

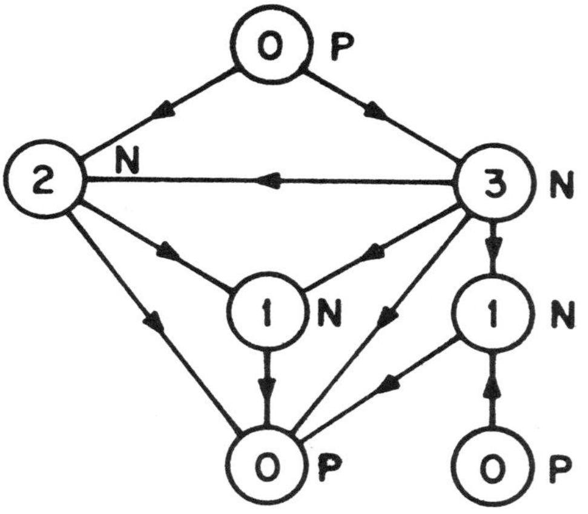

Figure 3. Random game-graph.

There is a simple algorithm for determining the N, P-pattern of any game-graph $G = (V, E)$. Define the set of *followers* of $u \in V$ by $F_G(u) = F(u) = \{v \in V\colon (u, v) \in E\}$. The definitions of N, P imply

$$u \in P \text{ if and only if } F(u) \subseteq N, \tag{1}$$

whereas

$$u \in N \text{ if and only if } F(u) \cap P \neq \emptyset. \tag{2}$$

Thus for normal play, all *leaves* (vertices with $F(u) = \emptyset$) are labeled P. Working backwards, using (1) and (2), one gets the N, P-labels in $O(|E|)$ steps (linear time). See Figs. 2 and 3 for some N, P-patterns. Since chess is a finite game, its game-graph can be constructed and its N, P, D-pattern can be computed, where D denotes *D*raw. (We shall see in §2.1 that the determination of the D-positions is also easy.) Hence our first conclusion is that chess is just as easy as Nim.

1.3. **Sums of Games**

Suppose that 2 players are given m disjoint combinatorial games. Each player at his turn selects a game and makes a move in it. Such a game is called the *sum* of the m component games, also known as *disjunctive compound* in the literature. As usual we consider normal play, i.e., the player first unable to move loses and his opponent wins.

Now put a number of tokens on some m vertices of Fig. 3. Each player at his turn selects a token and moves it, along a directed edge, to a neighboring vertex (whether occupied or not). Vertices may be multiply occupied. This game is equivalent to a game consisting of m copies of Fig. 3, with one token on each; in other words, it is a sum of m games.

The sum is a *Nim-like game* if its game-graph is finite and acyclic.

Place 3 tokens on the vertices labeled 1, 2, 3 of Fig. 2 and play according to the rules adopted for m tokens on Fig. 3. This sum is precisely the 3-pile Nim game described at the beginning. Also chess can be considered as a sort of compound, where several tokens are placed on an 8×8 board.

Put a token on each of the 2 vertices labeled 1 of Fig. 3, both of which are N-positions. This compound position is a P-position of the sum, because player II clearly wins. But a token on a vertex labeled 1 and on the vertex labeled 2, both of which are also N-positions, constitutes on N-position of the sum: player I can evidently win by moving $2 \to 1$. Thus 2 N's may sometimes be a P and sometimes an N-position of the sum. This shows that the N, P-tool is not strong enough to settle sums.

The trouble stems from the fact that playing a sum with m tokens on a graph (equivalently: m copies of a graph with a single token per copy) constitutes a game on a graph rather than a graph of a game! The latter is usually exponentially larger than the former as we shall now see.

1.4. Nim is Just as Hard as Chess

Let $G_i = (V_i, E_i)_{i=1}^{m}$ be m disjoint digraphs on which a sum is played, with precisely one token on each component graph. The game-graph $G = (\overline{V}, \overline{E})$ of the sum is the digraph with vertices $\overline{u} = (u_1, \ldots, u_m)$ for all $u_i \in V_i$ $(i = 1, \ldots, m)$. Let $\overline{v} = (v_1, \ldots, v_m) \in \overline{V}$. Then $(\overline{u}, \overline{v}) \in \overline{E}$ if and only if there exists k, $1 \le k \le n$, such that $v_k \in F(u_k)$ and $v_i = u_i$ for all $i \ne k$. Note that the vertex set $\overline{V}$ of a game-graph of a sum is the *product* of the vertex-sets of the m component digraphs.

For Nim with m piles, each of size at most n, the sum-graph has $\binom{m+n}{n}$ vertices, which is the number of m-combinations of $n+1$ distinct objects, each being repeated m times. For $n = m$, the sum-graph has $\binom{2n}{n} = \Theta(4^n/\sqrt{n})$ vertices. For $n \times n$ chess with $n^2/2$ pieces (which holds for $n = 8$), there are $\binom{n^2}{n^2/2} = \Theta(2^{n^2}/n)$ board configurations even if all chess pieces were identical. For both Nim and chess, the game-graph has thus exponentially many vertices.

If the best strategy for a game requires time which is nonpolynomial in the input size, we call the game *intractable*, consistent with the definition of an intractable problem, suggesting that a nonpolynomial computation is hard (see e.g. Garey and Johnson (1979)). Otherwise it is *tractable*. The strategy of an intractable game is normally just a variation of an exhaustive search through all game positions. Whether the conclusion now is that Nim is as hard as chess or chess as easy as Nim, we leave for the reader to judge! While you ponder this problem we shall go on to the next section, in which a tool for handling sums is developed.

1.5. **The Sprague-Grundy Function**

The tool which provides an efficient strategy for Nim and Nim-like games is the *Sprague-Grundy* function g (g-function for short). Let S be a finite set of nonnegative integers, $\overline{S} = \mathcal{Z}^0 - S$ its complement. Define mex $S = \min \overline{S}$ = least nonnegative integer not in S, where mex stands for *m*inimum *ex*cluded value. Then

$$g(u) = \text{mex } g\big(F(u)\big),$$

where for any set T and any function η on T, $\eta(T) = \big\{\eta(t) : t \in T\big\}$. For every finite acyclic digraph, g exists uniquely. See e.g. Berge (1985). Since mex $(\emptyset) = 0$, all leaves have g-value 0. Working backwards from the leaves, the g-values inside the vertices of Figs. 2 and 3 are obtained. Analogously to the N, P-labels, the g-function can be computed in $O(|E|)$ steps for every connected finite acyclic digraph $G(V, E)$. For normal play we have

$$N = \big\{u \in V : g(u) > 0\big\}, \qquad P = \big\{u \in V : g(u) = 0\big\}. \tag{3}$$

Indeed, for $u_1 \in V$ with $g(u_1) > 0$, player I, say, moves to a follower u_2 with $g(u_2) = 0$, from which player II moves to a follower u_3 which necessarily satisfies $g(u_3) > 0$, so eventually player I moves to a leaf, winning. Thus the g-function gives at least as much information as the N, P-labels. It actually gives more, as we shall now see. For $\overline{u} = (u_1, \ldots, u_m) \in \overline{V}$, where $G = (\overline{V}, \overline{E})$ is the game-graph of the sum, let

$$\sigma(\overline{u}) = \sum_{i=1}^{m}{}' g(u_i),$$

where the dash on $\sum$ indicates Nim-sum rather than ordinary sum. We claim that σ is the g-function of G, so (3) implies

$$N = \big\{\overline{u} \in \overline{V} : \sigma(\overline{u} > 0\big\}, \qquad P = \big\{\overline{u} \in \overline{V} : \sigma(\overline{u}) = 0\big\}. \tag{4}$$

The strategy of Nim is consistent with this claim, because Fig. 2 shows that $g(k) = k$, where k is size of the Nim-pile. Thus the rule given at the beginning of this article is a special case of the claim.

Rather than proving the claim, it will be convenient here to prove (4) directly. Suppose that $g(u_1), \ldots, g(u_m)$ are as shown

Table 2. Nim-sum 0.

$g(u_1) = 0$	1	0	1	$\cdots$	1
$g(u_2) = 1$	0	0	1	$\cdots$	0
$\vdots$					
$g(u_n) = 1$	1	0	0	$\cdots$	1
$\sigma(\overline{u}) = 0$	0	0	0	$\cdots$	0

Table 3. Positive Nim-sum.

$g(u_1) = 0$	0	0	0	$\cdots$	1
$g(u_2) = 1$	0	0	1	$\cdots$	0
$\vdots$					
$g(u_n) = 1$	1	0	0	$\cdots$	1
$\sigma(\overline{u}) = 0$	1	0	1	$\cdots$	0

in Table 2, and $\sigma(\overline{u}) = 0$. Any move from the position $\overline{u} = (u_1, \ldots, u_m)$ involves moving in precisely one u_i, say from u_1 to u_1'. The definition of g then implies $g(u_1') \neq g(u_1)$, hence in at least one column of Table 2 the parity of the number of 1-bits is reversed, and so $\sigma(\overline{u}') > 0$, where $\overline{u}' = (u_1', u_2, \ldots, u_n) \in F_G(\overline{u})$.

If $\sigma(\overline{u}) > 0$, there exists a leftmost column c_0 which contains the most significant 1-bit of $\sigma(\overline{u})$ (2nd column from left in Table 3). Hence in c_0 there is an odd number of 1's, so at least one. Without loss of generality, assume that this occurs in $g(u_n)$. Then put the c_0th bit of $g(u_n)$ to 0, and reverse the parity of the other bits of $g(u_n)$ precisely in those columns in which $\sigma(\overline{u})$ has a 1-bit. This results in a value t which clearly satisfies

$$0 \le t < g(u_n), \qquad t \oplus \sum_{i=1}^{n-1}{}' g(u_i) = 0,$$

where $\oplus$ denotes Nim-sum. The clinch is that there is indeed a move from u_n to some u_n' with $g(u_n') = t$, because $g(u_n)$ is the mex of the g-values of its followers, so for every $0 \le t < g(u_n)$ there is *some* $u_n' \in F(u_n)$ such that $g(u_n') = t$.

Thus the player in a position $\overline{u}_1$, with $\sigma(\overline{u}_1) > 0$ can always move to some $\overline{u}_2 \in F(\overline{u}_1)$ with $\sigma(\overline{u}_2) = 0$. If there is any move left, his opponent necessarily moves to some $\overline{u}_3 \in F(\overline{u}_2)$ with $\sigma(\overline{u}_3) > 0$, and so on. Since G is finite and acyclic, the player moving from $\overline{u}_1$, reaches some leaf $\overline{u}_p$ of G with $\sigma(\overline{u}_p) = 0$ in $O(mn)$ steps, where $n = \max_{1 \le i \le m} |V_i|$, proving (4).

1.6. **Some Nim-Like Games are Tractable**

We have seen that the g-value σ of the game-graph G of a sum can be computed in polynomial time in the standard sizes $\Theta\left(\sum_{i=1}^m |E_i|\right)$ of the disjoint component game-graphs $G_1, \ldots, G_m$. (Note that we don't have to construct G for computing σ.) It follows that Nim-like games (§1.3) are tractable in the sense of the definition at the end of §1.4.

Or does it? The keywords here are "standard input size" and "disjoint component game-graphs". These notions are not always strictly true, resulting in intriguing complications:

(a) The m component game-graphs containing up to n edges each are very specialized. Typically they may be described very succinctly, e.g. their combined input has size $\Theta(m \log n)$ rather than $\Theta(mn)$, which is the standard size. For bounded m, a strategy involving $O(mn)$ steps is then exponential.

(b) The m component digraphs $G_1, \ldots, G_m$ are not disjoint. Typically this happens when there is some dependence or interaction between the tokens. In other cases the component digraphs are not quite fixed. Both phenomena may prevent the Nim-sum σ from being the g-function of G. We may thus be back at square 1, being faced with the problem of an exponentially large game-graph.

Some ramifications of these observations will be taken up in §6.

2. Games With Draws

The above Sprague-Grundy theory was already known in the 30's — see Sprague (1935–36), Grundy (1939). Why don't games like checkers, Go and chess submit to this theory? Here and in subsequent chapters we mention some of the reasons for this.

2.1. Draw Positions

The more sophisticated games, even checkers, may contain dynamic tie positions, also called *draw* positions. These are non-leaf positions from which no player can force a win, but both have a non-losing move. The set of such positions will be denoted D. The definition implies

$$u \in D \quad \text{if and only if} \quad F(u) \cap P = \emptyset \quad \text{and} \quad F(u) \cap D \neq \emptyset. \quad (5)$$

Draws can occur only if the game-graph contains cycles or loops. A loop at u can be interpreted as a pass at u, i.e., the player passes up his turn at a move. Analogously to the acyclic case, there is an $O(|E|)$ algorithm for determining the N, P, D-labels, but they do not suffice for determining the N, P, D-labels of a sum-game without constructing the exponentially large sum-graph.

2.2. The Strong Murphy Law

We turn quite naturally to the g-function for help. The first unpleasant surprise is that the very decision problem whether g exists is NP-complete for cyclic digraphs, even rather simple ones: planar, degree ≤ 3, indegree and outdegree ≤ 2 at each vertex (Fraenkel (1981)). (See §5 for the notion of NP-completeness and

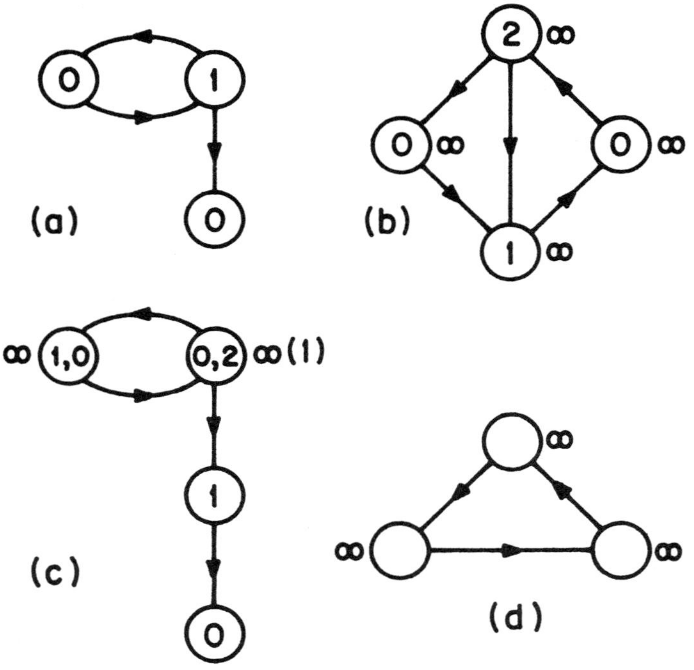

Figure 4.

(a) g exists uniquely and determines the strategy.
(b) g exists uniquely but does not determine the strategy.
(c) g exists non-uniquely and does not determine the strategy.
(d) g does not exist.

related complexity issues.) The second surprise is that everything that can go wrong with g will go wrong in the presence of cycles or loops.

(a) It may exist uniquely and determine the game's strategy, exactly as for Nim-like games (Fig. 4(a); g-values are depicted *inside* the vertices).

(b) It may exist uniquely but not determine the game's strategy. In Fig. 4(b) the g-values are unique. Since the digraph has no leaves, every position is a D-position, so the g-values do not determine the strategy.

(c) It may exist non-uniquely, in which case it cannot determine a strategy (this follows from the Fundamental Theorem—see §2.7). In Fig. 4(c), the 2 vertices on the cycle can assume one of 2 different g-values each.

(d) It may not exist at all (Fig. 4(d)).

2.3. Two Failures are Better than One — Sometimes!

A tool rendered NP-hard, such as the g-function on cyclic graphs, is next to useless. Here, in addition, it does not even do what it's supposed to do: determine the strategy. Bad as all this sounds, the second failure contains potential hope: if it does not determine the strategy, who cares about its NP-hardness? Rather than attempting to mend the shattered pieces, let's try and look for another tool that exists uniquely on all finite cyclic digraphs, can be constructed polynomially, and determines the strategy in all cases.

Fortunately, a nice tool with all these properties does exist. It is the Generalized Sprague-Grundy Function, or *γ-function* for short (Smith (1966), Fraenkel and Perl (1975), Fraenkel and Yesha (1986), ONAG* ch. 11, WW* ch. 12).

2.4. The Generalized Sprague-Grundy Function

Let $G = (V, E)$ be a finite digraph. A γ-function is a mapping $\gamma: V \rightarrow \mathcal{Z}^0 \cup \{\infty\}$, which satisfies conditions **A**, **B**, **C** detailed below. We denote by V^f the subset of V on which γ is finite, and $V^\infty = V - V^f$. Let $\gamma(F(u)) = \{\gamma(v) < \infty : v \in F(u)\}$. If $\gamma(u) = \infty$ and $\gamma(F(u)) = K$, we also write $\gamma(u) = \infty(K)$. A γ-function is often associated with a *counter function* $c: V^f \rightarrow \mathcal{Z}^0$. One motivation for c is that it permits the winner to choose the P-position he'll want to move to — one with minimal c-value usually.

---

* Here and below, ONAG stands for Conway (1976), and WW for Berlekamp, Conway and Guy (1982).

When on a cycle, moving merely to a P-follower does not guarantee a win, only non-losing (see e.g. the two 0-values on Fig. 4(a); only moving to the leaf is a win). Another motivation is that c permits proving properties of γ by induction on the c-value.

A. If $\gamma(u) < \infty$, then $\gamma(u) = \text{mex}\ \gamma\bigl(F(u)\bigr)$.

B. If $v \in F(u)$ and $\gamma(v) > \gamma(u)$, then there exists $w \in F(v)$ satisfying $\gamma(w) = \gamma(u)$ and $c(w) < c(u)$.

C. If $\gamma(u) = \infty$, then there is $v \in F(u)$ with $\gamma(v) = \infty(K)$ such that $\text{mex}\ \gamma\bigl(F(u)\bigr) \notin K$.

The least intuitive of these conditions is **C**. To make it more accessible, note that its converse follows from **A** and **B**. Indeed, suppose there is some $v \in F(u)$ with $\gamma(v) = \infty(K)$. If $\gamma(u) < \infty$, then there exists $w \in F(v)$ with $\gamma(w) = \gamma(u)$ (by **B**) $= \text{mex}\ \gamma\bigl(F(u)\bigr)$ (by **A**). In other words, $\text{mex}\ \gamma\bigl(F(u)\bigr) \in K$. Thus **C** could be reformulated in the stronger form: $\gamma(u) = \infty$ if and only if there is $v \in F(u)$ with $\gamma(v) = \infty(K)$ such that $\text{mex}\ \gamma\bigl(F(u)\bigr) \notin K$. But the present weaker form of **C** is preferable for a definition.

Analogously to the acyclic case, the γ-function gives at least as much information as the N, P, D-pattern, since

$$\begin{aligned} N &= \bigl\{u \in V : 0 < \gamma(u) < \infty\bigr\} \cup \bigl\{u \in V : \gamma(u) = \infty(K),\ 0 \in K\bigr\}, \\ P &= \bigl\{u \in V : \gamma(u) = 0\bigr\}, \quad D = \bigl\{u \in V : \gamma(u) = \infty(K),\ 0 \notin K\bigr\}, \end{aligned} \tag{6}$$

which is not hard to verify. Moreover, the γ-function exists uniquely for every finite digraph $G = (V, E)$.

For handling sums with possible draws, a generalized Nim-sum σ is defined as ordinary Nim-sum, but with 2 additional conditions:

$$a \oplus \infty(K) = \infty(K) \oplus a = \infty(K \oplus a),$$

where $K \oplus a = \{k \oplus a : k \in K\}$ (a any nonnegative integer), and $\infty(K) \oplus \infty(L) = \infty(\emptyset)$. Recall the definition of the game-graph $G = (\overline{V}, \overline{E})$ of the sum of m games given in §1.4. Analogously to the acyclic case, it turns out that also for the cyclic case σ (using the generalized Nim-sum) is the γ-function of G. Thus (6) implies for the sum,

$$\begin{aligned} N &= \{\overline{u} \in \overline{V}: 0 < \sigma(\overline{u}) < \infty\} \cup \{\overline{u} \in \overline{V}: \sigma(\overline{u}) = \infty(K),\ 0 \in K\}, \\ P &= \{\overline{u} \in \overline{V}: \sigma(\overline{u}) = 0\}, \quad D = \{\overline{u} \in \overline{V}: \sigma(\overline{u}) = \infty(K),\ 0 \notin K\}. \end{aligned} \tag{7}$$

This leads again to a polynomial time strategy in the standard input sizes of the m disjoint component game-graphs.

For example, a single token on the vertex with γ-value 1 in Fig. 4(a) is an N-position. The same holds if we adjoin to this token one on the vertex with γ-value $\infty(1)$ (Fig. 4(c)), since $\infty(1) \oplus 1 = \infty(1 \oplus 1) = \infty(0)$ (the unique winning move is $\infty(1) \to 1$). A single token on a vertex with γ-value 0 (Fig. 4(a)) is a P-position. Adjoining to this a token on $\infty(1)$ is a D-position, since $\infty(1) \oplus 0 = \infty(1)$. If another token on any vertex with γ-value ∞ is adjoined, the position is still a draw.

2.5. An Algorithm

How did we get the γ-values of Fig. 4? They can be computed using Algorithm GSG below. Initially a special symbol ν is attached to the label ℓ of every vertex u, where $\ell(u) = \nu$ means that u has no lable. We also let $V_\nu = \{u \in V : \ell(u) = \nu\}$, where $\ell(u)$ designates the label of u.

Algorithm GSG *for computing the Generalized Sprague-Grundy function for a given finite digraph* $G = (V, E)$.

1. (Initialize labels and counter.) Put $i \leftarrow 0$, $m \leftarrow 0$, $\ell(u) \leftarrow \nu$ for all $u \in V$.
2. (Label and counter.) If there exists $u \in V_\nu$ such that no follower of u is labeled i and every follower of u which is either unlabeled or labeled ∞ has a follower labeled i, then put $\ell(u) \leftarrow i$, $c(u) \leftarrow m$, $m \leftarrow m + 1$. Repeat 2.
3. (∞-label.) For every $u \in V_\nu$ which has no follower labeled i, put $\ell(u) \leftarrow \infty$.

4. (Increase label.) If $V_\nu \neq \emptyset$, put $i \leftarrow i+1$ and return to 2; otherwise end.

Note that since the algorithm has to be capable of labeling a first vertex on a cycle, it cannot be expected that a vertex will be labeled only after all its followers have been labeled, as for g. In step 2, which is the heart of Algorithm GSG, a vertex u is indeed labeled i at a stage where u may yet have unlabeled followers. However, since each of them has to have a follower labeled i, none of the followers of such u can ever be labeled i if **A** is to hold. This complication explains the $O(|E||V|)$ complexity of Algorithm GSG for any connected finite digraph $G = (V, E)$, which is higher than that for computing g, but still polynomial.

2.6. Learn to Beat the Masters

Consider the game displayed in Fig. 5. An initial position might be tokens on vertices A, B, C, D. Each player at his turn selects a token and moves it to an adjacent vertex (occupied or not) along a directed edge. The player first unable to move loses (when all tokens are on L), his opponent wins. If there is no last move, the outcome is a draw. Is the given position an N, P or D-position? John Conway has asked how long it would take a professional chess-player to answer this question.

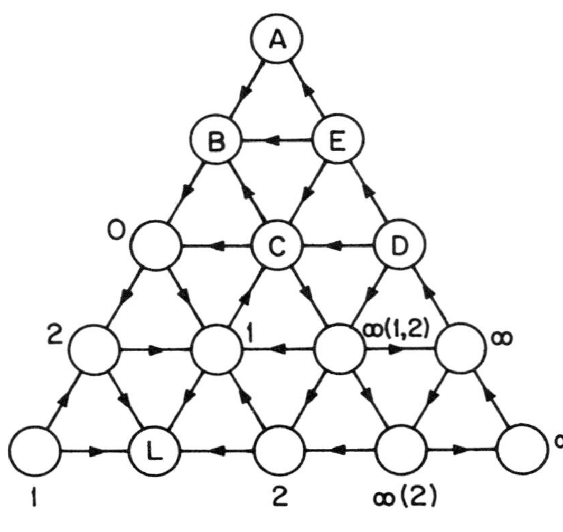

Figure 5. Learn to beat today's chess champion.

It turns out that the given position p is an N-position. But there is only one way to win: $D \to E$. This follows from $\gamma(A) = 0$, $\gamma(B) = 1$, $\gamma(C) = 2$, $\gamma(D) = \infty(2,3)$, $\gamma(E) = 3$, and from (7): $\gamma(p) = 0 \oplus 1 \oplus 2 \oplus \infty(2,3) = 3 \oplus \infty(2,3) = \infty(0,1)$. The γ-values of the other circles appear on Fig. 5, and of course the leaf L satisfies $\gamma(L) = 0$. These values enable the winner to realize his win even without a counter function. For the case of 2 *human* players playing on digraphs of moderate size, no counter function is needed: humans — whose perception lies somewhere between that of deterministic and nondeterministic machines! — learn quickly to "guess" a follower with specified γ-value of decreasing counter function value, if not of minimal value, even though the counter is not given explicitly. But things are trickier for annihilation games; this is hinted at in §3.6 below.

Having learned to win in this game, you can now tackle that of Fig. 6, which has the same rules but is somewhat larger. Do you agree that the indicated position is also an N-position, and that the unique winning move is to move horizontally from E? And what is the nature of the same position but without the token on D?

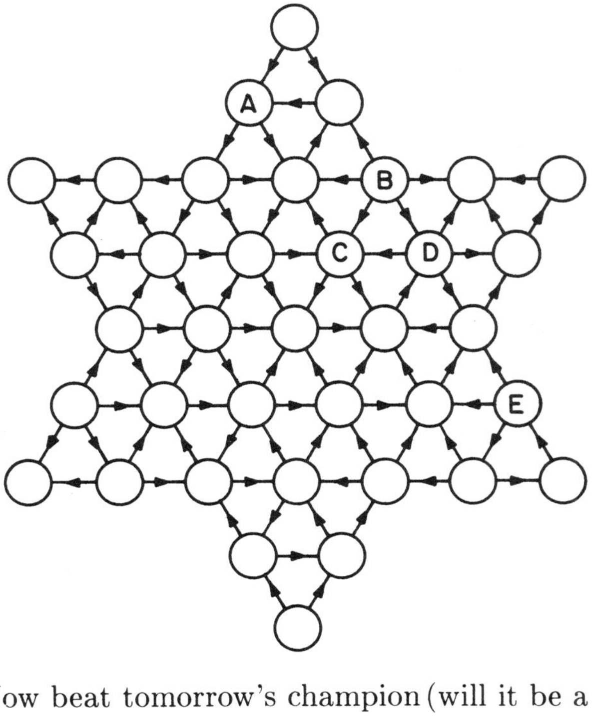

Figure 6. Now beat tomorrow's champion (will it be a computer?).

2.7. Fundamental Theorem of Combinatorial Game Theory

We close this chapter with an answer to a question that may have been (in fact: should have been) nagging deep inside you from the very beginning of this story: Is it clear that every position in every game has a unique N, P or D-label? Perhaps some positions are undetermined or have 2 or 3 labels? When playing a game with a friend, haven't both of you made some opening simplifying moves to get to positions from which the outcome is determined, implying that the initial position does not determine the outcome?

Analogously to the Fundamental Theorem of Algebra and the Fundamental Theorem of Arithmetic, we state the following result.

Fundamental Theorem of Combinatorial Game Theory. *Let Γ be a 2-person game with perfect information and no chance moves, whose game-graph may be infinite. Then for every position of Γ there are precisely two possibilities*: (i) *There exists a winning move for precisely one of the two players.* (ii) *There is a winning move for neither player, but both can maintain a draw.*

The main argument of the proof is this: Suppose that u_1 is a position of Γ which has been labeled neither N nor P. Then $F(u_1) \not\subseteq N$, otherwise $u_1 \in P$ by (1). Moreover, $F(u_1) \cap P = \emptyset$, otherwise $u_1 \in N$ by (2). Assume $u_1 \notin D$. Then (5) implies $F(u_1) \cap D = \emptyset$. Thus there is $u_2 \in F(u_1)$ such that also u_2 is unlabeled. Hence there exists an infinite path $u_1, u_2, \ldots$ of unlabeled (not necessarily distinct) vertices with $u_{i+1} \in F(u_i)$. An optimal strategy is to move $u_i \to u_{i+1}$ for all $i \geq 1$, since the only labeled followers of u_j, if any, are N-positions for all $j \geq 1$. This implies $u_i \in D$ for all $i \geq 1$, a contradiction. Hence every vertex is in fact labeled.

3. Annihilation Games

A second reason why chess-like games are not as tame as Nim-like games is due to interaction among tokens, such as various capture rules. A particular case is *annihilation*, suggested by John Conway.

3.1. The Game-Graph ...

An annihilation game is played on a finite digraph $G = (V, E)$ by initially placing at most one token on every vertex. Each player at his turn moves a token along a directed edge, say from u to v for $(u, v) \in E$. If the move $u \to v$ results in 2 tokens on v (v had been occupied), then both tokens are removed from the game ("annihilation" move). See Fraenkel (1974), Fraenkel and Yesha (1976, 1982), and Fraenkel, Tassa and Yesha (1978). As usual we consider normal play, but we note that misère play of annihilation games has been considered by Ferguson (1984).

If G is acyclic, the strategy for the annihilation game on G is the same as for the sum-game, except that a play of the former may have less moves. This assertion stems from the fact that 2 tokens coinciding on the same vertex u in the sum-game satisfy $g(u) \oplus g(u) = 0$, so the strategy is invariant under the removal of both tokens. The situation is quite different if G has cycles, however. Consider for example a position consisting of two tokens on vertices A and B in Fig. 7. As an ordinary sum, this is clearly a D-position, since there is no leaf. But as an annihilation game it is a P-position: player I moves, say $B \to C$, and then player II wins making the annihilation move $C \to A$. However, the position (A, B) in Fig. 8 is a D-position for both the ordinary sum and the annihilation games. As in §2 we permit G to have loops.

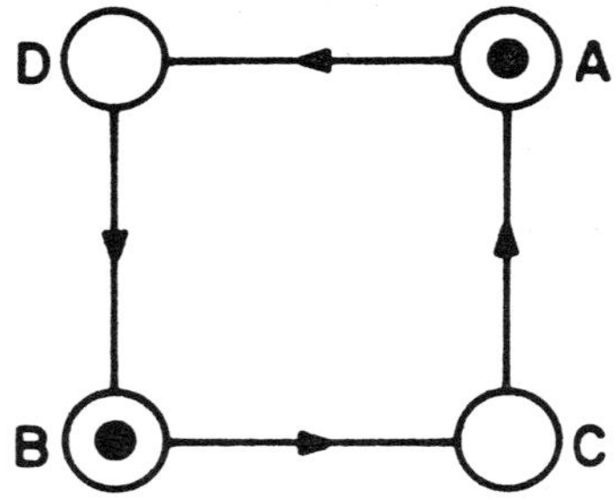

Figure 7. A draw-position in a non-annihilation game, which is a P-position in the annihilation-version.

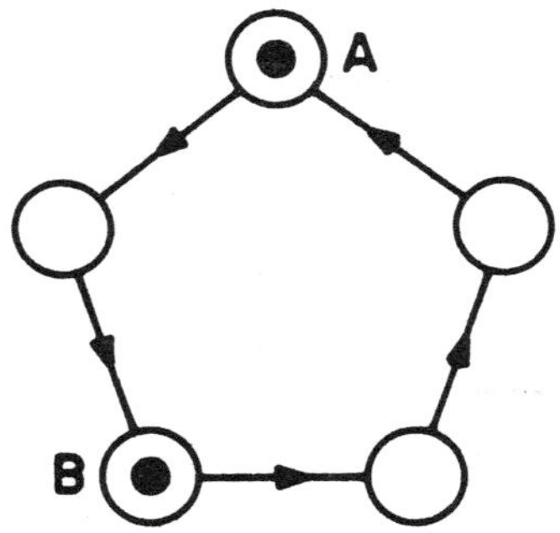

Figure 8. A draw-position in both the annihilation and the non-annihilation game.

3.2. ... **Is Exponentially Large**

Order V in some way, say

$$V = (z_1, \dots, z_n).$$

Let $\overline{G} = (\overline{V}, \overline{E})$ denote the game-graph of the annihilation game played on $G = (V, E)$. Then any position in $\overline{G}$ can be described in the form $\overline{u} = (u^1, \dots, u^n) \in \overline{V}$ over the field $GF(2)$, where $u^k = 1$ if and only if there is a token on z_k. In particular, the vector $\Phi = (0,\ldots,0)$ is a leaf in $\overline{V}$ and $|\overline{V}| = 2^n$. Thus $\overline{V}$ has exponential size.

Define unit vectors $\overline{z}_i = (z_i^1,\ldots,z_i^n)$ with $z_i^j = 1$ if and only if $i = j$. Then $(\overline{u},\overline{v}) \in \overline{E}$ if and only if there exist k and ℓ such that $u^k = 1$, $z_\ell \in F_G(z_k)$ and $\overline{v} = \overline{u} \oplus \overline{z}_k \oplus \overline{z}_\ell$, where $\oplus$ denotes Nim-sum (addition over $GF(2)$) as before. Note that the move $\overline{u} \to \overline{v}$ involves annihilation if and only if $u^\ell = 1$ and $k \neq \ell$ (for $k = \ell$ the move is along a loop). The story of the next 2 subsections is how to compute γ on $\overline{G}$ polynomially despite the large size of $\overline{G}$. We use the notation

$$\overline{V}_i = \{\overline{u} \in \overline{V}^f : \gamma(\overline{u}) = i\} \qquad (i \geq 0).$$

3.3. **Vector Spaces, Homomorphisms, Quotient Spaces**

Note that $\overline{V}$ is a vector space over $GF(2)$, satisfying $1 \cdot \overline{u} = \overline{u}$ for all $\overline{u} \in \overline{V}$. The first key observation is that $\gamma(\overline{u}_1 \oplus \overline{u}_2) = \gamma(\overline{u}_1) \oplus \gamma(\overline{u}_2)$ if either $\gamma(\overline{u}_1) < \infty$ or $\gamma(\overline{u}_2) < \infty$ (it is not generally true if $\gamma(\overline{u}_1) = \gamma(\overline{u}_2) = \infty$). This observation, which does not seem too easy to prove, leads to the following conclusions:

(i) $\overline{V}^f$ and $\overline{V}_0$ are subspaces of $\overline{V}$.

(ii) γ is a homomorphism from $\overline{V}^f$ onto $GF(2)^t$ (t-tuples over $GF(2)$) for some $t \in \mathcal{Z}^0$ with kernel $\overline{V}_0$ and quotient space $\overline{V}^f/\overline{V}_0 = \{\overline{V}_i : 0 \leq i < 2^t\}$, $\dim(\overline{V}^f) = m + t$, where $m = \dim(\overline{V}_0)$.

Thus $\overline{V}^f$ has a very pleasing structure. In particular, for every i, $0 \leq i \leq 2^t - 1$, there is some $\overline{u} \in \overline{V}^f$ with $\gamma(\overline{u}) = i$, and all such $\overline{u}$ form the i-th coset of $\overline{V}_0$.

We wish to extend the homomorphism $\gamma: \overline{V}^f \to GF(2)^t$ to a homomorphism $\overline{\gamma}$ on the entire space $\overline{V}$. Since any $\overline{u} \in \overline{V}$ can be written as a linear combination of the unit vectors, i.e., $\overline{u} = \sum_{i=1}^{\prime n} \varepsilon_i \overline{z}_i$ ($\varepsilon_i \in \{0,1\}$, $1 \leq i \leq n$), but some or all of the γ-values of the $\overline{z}_i$ may be ∞, the extended homomorphism will then permit to compute

$$\overline{\gamma}(\overline{u}) = \sum_{i=1}^{n}{}' \varepsilon_i \overline{\gamma}(\overline{z}_i), \tag{8}$$

such that $\overline{u} \in \overline{V}^f$ if and only if $\overline{\gamma}(\overline{u})$ is essentially $\gamma(\overline{u})$, modulo the extension.

Such an extended homomorphism $\overline{\gamma}: \overline{V} \to GF(2)^{n-m}$ does exist; it has the property that $\overline{u} \in \overline{V}^f$ if and only if the leftmost $n - m - t$ bits of $\overline{\gamma}(\overline{u})$ are 0.

3.4. An $O(n^6)$ algorithm for γ

Let $\overline{V}^{[m]} = \{\overline{u} \in \overline{V}: \sum_{k=1}^{n} u^k \le m\}$. The second key observation is that $\overline{V}^f$ is spanned by a subset of $\overline{V}^{[4]}$, i.e., $\overline{V}^f$ is the linear span of a subset of all vectors with at most four 1-bits only! This subspace has size $O(n^4)$. The rest is standard linear algebra over $GF(2)$: Computing a basis for $\overline{V}^f$ and $\overline{V}_0$ can be done by transforming the $n \times O(n^4)$ matrix M of $\overline{V}^{[4]}$ into row-echelon form ($O(n^6)$ steps). By adjoining the identity matrix I_n to the end of M, the elementary row operations performed on the augmented matrix will yield the values $\overline{\gamma}(\overline{z}_i)$, so $\overline{\gamma}(\overline{u})$ can be computed easily for all $\overline{u} \in \overline{V}$ by (8). Then $\gamma(\overline{u}) = \overline{\gamma}(\overline{u})$ if the $n - m - t$ leftmost bits of $\overline{\gamma}(\overline{u})$ are 0; otherwise $\gamma(\overline{u}) = \infty$.

Polynomial algorithms are usually of the order of n^3 or less. The above indicated algorithm is unusual in the high exponent n^6. Though it is only an upper bound, it does not seem easy to get a small upper bound for a general digraph.

3.5. Complexity of Winning Strategies

Computing a strategy of play for a given position u of a game usually consists of 2 steps:

(i) Finding whether u is in N, P or D.

(ii) If u is in N or in D, computing an optimal next move.

When we say that a strategy is tractable, we mean that both steps are polynomial. It is intractable if either of these steps is not polynomial. Following Kalmár (1928) and Smith (1966), we define a strategy in the *narrow sense* to be a strategy which depends only on the current position. A strategy in the *wide sense* depends on the previous positions as well. Clearly, every combinatorial game has a strategy in the narrow sense. For Nim-like games, the winner moves from his present N-position to *any* P-follower, and the loser has to move to an N-follower. After a finite number of moves, the winner reaches a leaf. The situation is similar in games with possible draw-positions without annihilation. However, the winner can't just move to any P-follower; he has to move to a P-follower with smallest or at least smaller counter function value in order to avoid cycling indefinitely.

The existence of a polynomial algorithm for computing the N, P, D-pattern for an annihilation game does not imply a polynomial strategy of play, however: we computed the finite γ-values of $\overline{G}$ by restricting attention to a subset of $\overline{V}^{[4]}$, so we don't have a counter function on all of $\overline{V}^f$ to prevent possible cycling. Also, the counter function of G does not carry over to $\overline{G}$; in fact, G does not have any counter function if $\gamma(u) = \infty$ for all $u \in V$, such as in Figs. 7 and 8. It turns out that a polynomial strategy in the broad sense can be computed in $O(n^6)$ steps, such that the winner can win in $O(n^5)$ moves. A strategy in the *broad sense* is a strategy which depends on the present position and on all possible previous positions, not only those previous positions actually encountered. It does not yet seem to be known whether there exists a strategy in the narrow sense which is polynomial in n.

3.6. Innocent Marbles

The Innocent Marble game is displayed in Fig. 9. Place an even number of indistinguishable "marbles" on distinct vertices of each of the 5 components. Gill and Jean play by alternately selecting one of the 10 marbles, moving it to a neighboring vertex u, along

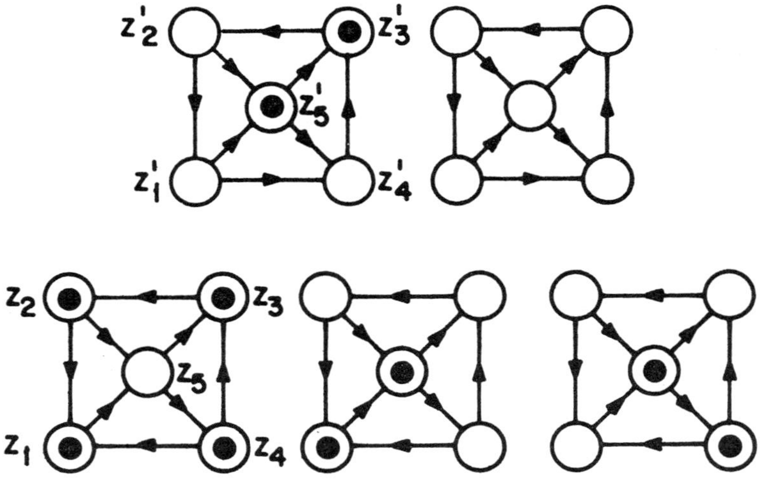

Figure 9. The Innocent Marble game.

a directed edge. If u was occupied, both marbles are removed from the game. The player first unable to move loses and the opponent wins, collecting all the marbles. If there is no last move, the outcome is a draw. (The outcome is trivially a draw if there is an odd number of marbles.) Note that the 3 components on the bottom row are isomorphic; also the 2 components on top are isomorphic, and the only difference between a bottom and a top component is the direction of the bottom edges.

It turns out that the position depicted in Fig. 9 is an N-position, and there is a unique winning move.

Question 1. Which is it?

It may be useful to consider easier questions first. Suppose e.g. that we have only 4 marbles, namely on z_4 and z_5 on 2 (bottom) components. Without doing any computation, it is evident that player II can at least draw, simply by imitating on one component what player I does on the other.

Question 2. Can player II do better and win?

Question 3. What is the nature of the position consisting of 4 marbles, on z_4, z_5, z_4', z_5'?

Using the above indicated algorithm, it is easy to compute the values $\overline{\gamma}(\overline{z}_1) = \overline{\gamma}(\overline{z}_3) = 5$, $\overline{\gamma}(\overline{z}_2) = \overline{\gamma}(\overline{z}_4) = 4$, $\overline{\gamma}(\overline{z}_5) = 7$, and $n = 5$, $m = t = 2$. Here and below we use a decimal encoding for $GF(2)^t$, i.e., $\overline{\gamma}(\overline{z}_3) = 5$ stands for the binary vector 101. Now $\overline{u} \in \overline{V}^f$ if and only if the $n - m - t = 1$st most significant bit of $\overline{\gamma}(\overline{u})$ is 0. Thus,

$$\overline{\gamma}(\overline{z}_4 \oplus \overline{z}_5) = \overline{\gamma}(24) = \overline{\gamma}(\overline{z}_4) \oplus \overline{\gamma}(\overline{z}_5) = 4 \oplus 7 = 3 = \gamma(24).$$

Hence the answer to Question 2 is that player II can indeed win! We did not give in §3.5 the method to compute the next move; we only discussed its complexity. But with this small graph and the knowledge of the $\overline{\gamma}(\overline{z}_i)$, the reader will have no difficulty in actually winning as second player against his favorite adversary. It is also easy to see that the $\overline{\gamma}$-values of the 8 marbles on the bottom 3 components are $0 \oplus 2 \oplus 3 = 1$, which is thus also the γ-value.

For the top components we get $\overline{\gamma}(\overline{z}'_1) = \overline{\gamma}(\overline{z}'_3) = 3$, $\overline{\gamma}(\overline{z}'_2) = \overline{\gamma}(\overline{z}'_4) = 2$, $\overline{\gamma}(\overline{z}'_5) = 4$, and $n = 5$, $m = 2$, $t = 1$. Then $\overline{u} \in \overline{V}^f$ if and only if the $n - m - t = 2$ most significant bits of $\overline{\gamma}(\overline{u})$ are 0. Thus

$$\overline{\gamma}(\overline{z}'_4 \oplus \overline{z}'_5) = \overline{\gamma}(24) = \overline{\gamma}(\overline{z}'_4) \oplus \overline{\gamma}(\overline{z}'_5) = 2 \oplus 4 = 6,$$

hence $\gamma(24) = \infty$. Since $\overline{\gamma}(\overline{z}'_3 \oplus \overline{z}'_4) = \overline{\gamma}(12) = 3 \oplus 2 = 1 = \gamma(12)$, we actually have $\gamma(24) = \infty(0,1)$. Thus for Question 3 we get the value $2 \oplus \infty(0,1) = \infty(2,3)$, so the position is a draw. It can be maintained by moving with the token on either z_4 or z'_4, but not onto z_5 or z'_5 respectively. For the 10 marbles of Question 1 we have the γ-value $1 \oplus \infty(0,1) = \infty(0,1)$. Hence the unique winning move is $z'_5 \to z'_4$.

3.7. **Worlds in Collision; Particles and Antiparticles**

After mastering Innocent Marbles, we may like to tackle Worlds in Collision (Fig. 10), in which "worlds" or "stars" are placed initially in certain distinct "space stations" (vertices). Each of two

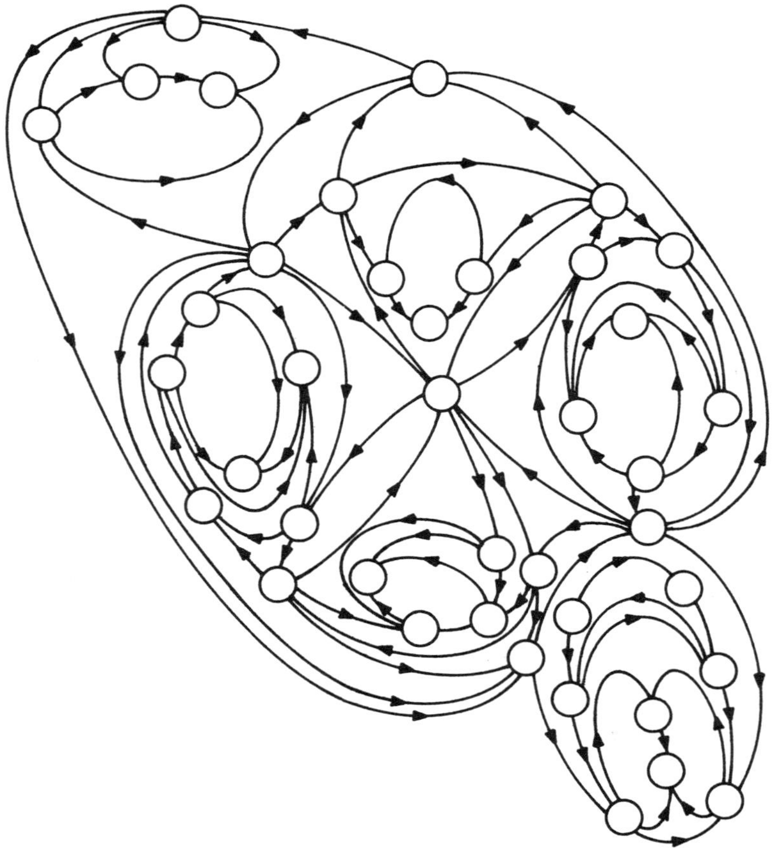

Figure 10. Worlds in Collision.

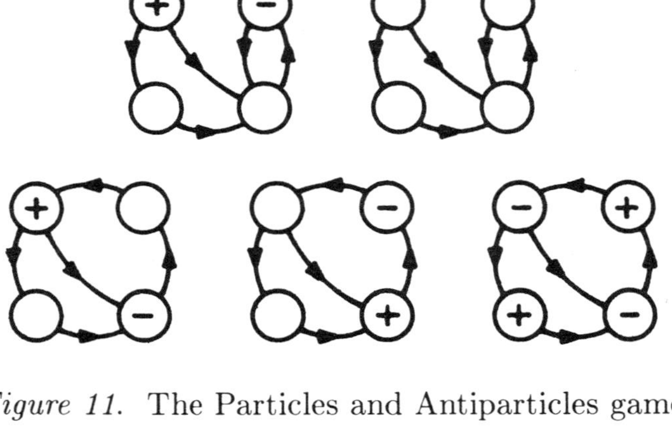

Figure 11. The Particles and Antiparticles game.

"superspacemen" can streak a star from a station to a neighboring station along a directed edge. Two colliding stars explode and vanish in a cloud of interstellar dust. In contrast to marbles, which get removed only in pairs, stars can disappear also singly in the "black holes" (leaves), where they vanish ("falling stars"), since they cannot reemerge from a black hole to streak again across the firmament!

A variation is the Particles and Antiparticles game played on the 5 orbits of Fig. 11. Place i electrons $(-)$ and i positrons $(+)$ on distinct vertices of each orbit ($i = 0, 1$ or 2). A typical arrangement is shown in the figure. Two physicists play the game by moving alternately. Each physicist at his turn selects a particle and moves it to an adjacent vertex u along a directed edge, provided that u is either unoccupied or occupied by a particle of the opposite type. In the latter case, of course, both particles get annihilated, as physicists tell us particles and antiparticles do. The lose-win-draw outcomes are defined as usual.

There is an easy strategy for this game played on the given simple graph or on other simple graphs. For example, the given position, in case you did not yet guess it, is a draw. But a careless move on your side can easily make you lose! However, we don't know a general strategy for this game played on an arbitrary digraph with n vertices, since it is not quite covered by the theory of annihilation games. If there is a strategy which is polynomial in n, it may not be all that simple: A special case, when all particles are of the same type — say electrons — is a generalization of the game *Welter*, which has an exotic strategy (see §6.2 and problems 35–37 in the chapter "Unsolved Problems in Combinatorial Games" in this volume).

4. Partizan Games

Another reason why the theory of Nim-like games is inadequate to handle chess-like games is the partizanship of the latter.

A game is *impartial* if both players have the same set of moves for all game positions. Otherwise the game is *partizan*. Nim-like games are impartial. Chess-like games are partizan, because if Gill plays the black pieces, Jean will not let him touch the white pieces.

In this section we shall denote a partizan game simply by the word game. The following two inductive definitions are due to Conway (1977, 1978(*Math. Mag.*), ONAG). See also WW.

(i) If M^L, M^R are any two sets of games, there exists a game $M = \{M^L|M^R\}$. All games are constructed in this way.

(ii) If LE, RI are any two sets of numbers and no member of LE is $\geq$ any member of RI, then there exists a number $\{LE \mid RI\}$. All numbers are constructed in this way.

Thus the numbers constitute a subclass of the class of games.

The first games and numbers are created by putting $M^L = M^R = LE = RI = \emptyset$. Some samples are given in Fig. 12, where

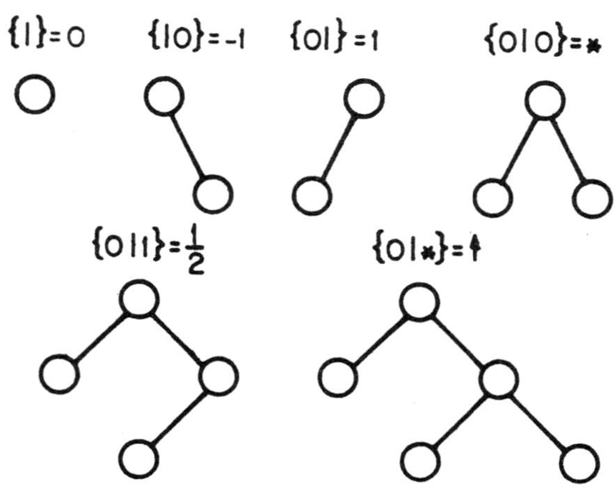

Figure 12. A few early partizan games.

L (Left) plays to the south-west and R (Right) to the south-east. If, as usual, the player first unable to move is the loser and his opponent the winner, then the examples suggest:

$M > 0$ if L can win,

$M < 0$ if R can win,

$M = 0$ if player II can win,

$M \| 0$ if player I can win (for example in $*$),

and we shall in fact define $>$, $<$, $=$, $\|$ by these conditions. The relations can be combined as follows:

$M \geq 0$ if L can win as player II,

$M \leq 0$ if R can win as player II,

$M \mathrel{|\!\rhd} 0$ if L can win as player I,

$M \mathrel{\lhd\!|} 0$ if R can win as player I.

Alternative inductive definitions of $\leq$, $\|$ can be given. Let $x = \{x^L \mid x^R\}$. Then

$x \leq y$ if and only if $x^L \mathrel{\lhd\!|} y$ for all x^L and $x \mathrel{\lhd\!|} y^R$ for all y^R,

$x = y$ if and only if $x \leq y$ and $y \leq x$,

$x \| y$ (x is *fuzzy* with y) if and only if $x \not\leq y$ and $y \not\leq x$,

as well as a consistency proof of both definitions. This enables proving many properties of games in a simple manner. For example, define $-M = \{-M^R | -M^L\}$. Then $G - G = 0$ (player II can win in $G - G$ by "imitating" the moves of player I in the other game). Also:

$$x \leq y \mathrel{\lhd\!|} z \quad \text{or} \quad x \mathrel{\lhd\!|} y \leq z \Rightarrow x \mathrel{\lhd\!|} z$$

(consider the game $z - x = (z - y) + (y - x)$, in which L can win as first player).

$$x^L \mathrel{\lhd\!|} x \mathrel{\lhd\!|} x^R$$

(clearly L can win as player I in $x - x^L$ and in $x^R - x$).

If x is a number, then $x^L < x < x^R$.

Most important is that a sum of games simply becomes a *sum*, defined by:

$$M + H = \{M^L + H, M + H^L \mid M^R + H, M + H^R\}.$$

Consider for example the game of Domineering (ONAG ch. 10, WW ch. 5), played with dominoes each of which covers precisely two squares of an $n \times n$ chessboard. L tiles vertically, R horizontally. After playing Domineering for a while, the board may break up into several parts, whence the game becomes a sum of these parts. The values of several small configurations are given in Fig. 13.

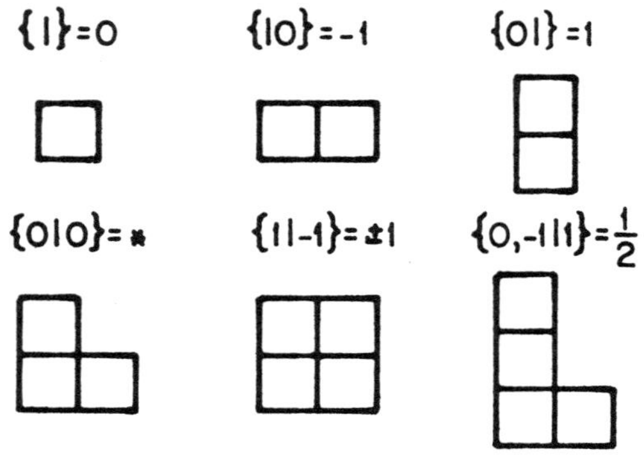

Figure 13. A few values of the Domineering game.

Since the relation of the value of a game M with 0 determines the strategy for playing it, computing the value of M is fundamental to the theory. In this direction the *Simplicity Theorem* is helpful: Let $x = \{x^L \mid x^R\}$ be any game. If there exists a *number* z such that $x^L \lhd z \rhd x^R$ and no option of z satisfies this relation, then $x = z$. Thus $\{-1 \mid 99\} = 0$. There also exist algorithms for computing values associated with games which are not numbers. Partizan games with possible draws are discussed in Li (1976), Conway (1978(*Ann. Discrete Math.*)), Shaki (1979), Fraenkel and Tassa (1982), WW ch. 11.

5. Provably and Conditionally Intractable Games

As we gradually increase the complexities of games from Nim-like games to chess-like games, it becomes increasingly harder to find polynomial strategies. The failure to find a polynomial strategy for a family of games may be due to our stupidity. But it may also be because there is no polynomial strategy!

5.1. Provably Intractable Games

To show that a family of games is intractable seems to be difficult, but some beginnings in this direction were made by Stockmeyer and Chandra (1979), who proved intractability of certain classes of games played on Boolean formulas. In Fraenkel and Lichtenstein (1981), a "real" game was proved intractable, namely chess on an $n \times n$ board. It was shown that for any reasonable generalization of chess to an $n \times n$ board, there exist constants $c > 1$, $d > 0$ and infinitely many chess-positions p such that any algorithm whether White (Black) can win from p requires at least $c^{|p|^d}$ time to compute, where $|p|$ is the size of p. Robson (1983, 1984) proved similar results for $n \times n$ Go and checkers.

There are not that many problems which have been proved intractable. Many problems are easily shown to be polynomial, simply by producing a polynomial algorithm, such as sorting n numbers, for which there are algorithms of complexity $O(n \log n)$. But for many other problems, both tractability and intractability seem difficult to establish. For these we do a next best thing, i.e., prove *conditional* intractability. Such results, *suggesting* intractability, establish that finding a strategy for certain classes of games is "just as hard" as a few hundred other problems for none of which a polynomial solution has been found, but none of which has been shown to be intractable either.

5.2. Some Complexity Classes

Let us define informally, and rather simplistically, a Deterministic Turing Machine (DM) to be a modern-day digital computer, and a Nondeterministic Turing Machine (NM) to be a DM in which the "next-move" function is multiple-valued. Thus an NM may traverse the branches of a rooted tree "in parallel", from the root to the leaves. A DM can do this only branch after branch. The set of problems which can be solved in polynomial time on a DM and on an NM is denoted by P and NP respectively. Clearly $P \subseteq NP$. It is an unsolved problem whether $P = NP$ or not. However, there are *complete* problems in NP: A problem Q is NP-*complete* if: (i) $Q \in NP$; (ii) existence of a polynomial algorithm for Q implies that $NP = P$. A problem is NP-*hard* if it satisfies (ii), but (i) does not necessarily hold. Any NP-hard problem is at least as hard as any NP-complete problem. The NP-complete problems are the "hardest" in NP: they lie in $NP - P$ if $NP \neq P$.

Another family of interest, called Pspace, consists of all problems which can be solved on an NM by using a polynomial amount of memory cells. Since in a polynomial amount of time only a polynomial amount of memory can be consulted, $NP \subseteq P$space. Again it is not known whether the inclusion is proper or not. Any Pspace-hard problem is at least as hard as any Pspace-complete problem (defined analogously to NP-hard and NP-complete), and the latter is at least as hard as an NP-complete problem. For further details see Garey and Johnson (1979). The class of problems which can be solved on an NM in exponential time is denoted Extime, and we have Pspace $\subseteq E$xtime. Problems complete in Extime, such as $n \times n$ chess, are intractable.

5.3. Some Games are Harder than Packing a Knapsack

Existential problems tend to lie in NP. These are problems involving a single existential quantifier. Examples are the Traveling Salesman problem (TS): Is there a round tour of n cities of cost $\leq C$?; and the Hamiltonian Cycle problem (HC): Is there a circuit containing every vertex of a given graph exactly once? These 2 problems are in fact NP-complete. However, a claim that player I can win in a game has the form "$\exists\forall\exists\forall \ldots$ moves : player I wins", i.e., there exists a move of player I such that for every move of player II there is a move of player I... such that player I can win. This is not a path in a tree (as for TS or HC), and it is not clear whether it is in *NP*. But many games lie in *P*space or *E*xtime. Typically games are Pspace-hard or Pspace-complete. Thus the frivolous activity of playing games is associated with higher complexities than TS, HC, or the knapsack problem, which are only NP-complete!

5.4. Conditionally Intractable Games

Many games have been proved Pspace-complete by Schaefer (1978). Since our complexity measures are asymptotic, they are not applicable to, say, checkers on an 8×8 board or Go on a 19×19 board. But on $n \times n$ boards, both are Pspace-hard (Fraenkel, Garey, Johnson, Schaefer and Yesha (1978), Lichtenstein and Sipser (1980 (1978))). Earlier, Even and Tarjan (1976) showed that Hex on an arbitrary graph is Pspace-complete. For Hex on an $n \times n$ board this was shown by Reisch (1981). Berlekamp has shown that Redwood Furniture, a special case of Left-Right Hackenbush [ONAG, WW], is NP-hard. Robertson and Munro (1978) have shown that Generalized Instant Insanity is NP-complete. (Of course a game with such a name cannot belong to our respectable family of combinatorial games! It is indeed a solitaire (one-player) game.) These and many more hardness results are summarized in Garey and Johnson (1979) and Johnson (1983).

5.5. Boundary Between Tractable and Intractable Games

It turns out that the annihilation games shown to be polynomial in §3, are only just so. Any perturbation seems to lead to NP-hard, Pspace-hard or Pspace-complete games. For example, suppose that tokens of r different types are placed on distinct vertices of a finite digraph $G = (V, E)$, whose edges are distributed into r subsets $E_1, \ldots, E_r$ (not necessarily disjoint), such that $\bigcup_{i=1}^{r} E_i = E$. A move consists of selecting a token of type i on some $u \in V$ and moving it to some $v \in F_i(u)$, where $F_i(u) = \{v \in V: (u, v) \in E_i\}$ $(1 \leq i \leq r)$. If v is occupied by a token of any type, the 2 tokens annihilate each other and are removed from the digraph. The game is impartial, and the case $r = 1$ is the annihilation game of §3. But for $r = 2$, $E_1 \cap E_2 = \emptyset$ and G an acyclic digraph, the decision problem whether any given game-position is an N-position is Pspace-complete. (On a cyclic digraph this problem is Pspace-hard.)

A partizan version of this game is obtained by using tokens of one type only, and $r = 2$. Player i at his turn moves a token from some $u \in V$ to some $v \in F_i(u)$, where the notation $F_i(u)$ is as in the previous game ($i \in \{1, 2\}$). If v is occupied, both tokens get annihilated. The question whether a given position is a P-position is NP-hard even if G is acyclic and bipartite. For these and other results on capturing or blocking tokens and marking or removing nodes, see Fraenkel and Yesha (1979), and Fraenkel and Goldschmidt (1987).

Regarding partizan games, Morris (1981) has shown that the decision problem whether a given sum of rather simple partizan games is a win for L, is Pspace-complete.

5.6. **Pathological Games**

Much of the theory outlined in chapters 1–3 can be generalized to certain classes of infinite digraphs. A *path* in a digraph $G = (V, E)$ is any sequence of vertices $u_1, u_2, \ldots$, not necessarily distinct, such that $(u_i, u_{i+1}) \in E$ ($i \geq 1$). Edges may be repeated. A *trail* is a path with distinct vertices. The *length* of a path or trail is the number of its edges, counting repetitions (for paths). If every path in G has finite length, G is called *path-finite*, and if every trail has finite length, G is called *trail-finite*. Much of the results of chapter 1 can be generalized to path-finite graphs, and much of chapter 2 to trail-finite graphs.

For the existence of the g- and γ-functions, an additional requirement is sufficient. A digraph $G = (V, E)$ is *locally path-bounded*, if for every vertex u_i there is a bound $b_i \in \mathcal{Z}^0$ such that the length of every (directed) path emanating from u_i does not exceed b_i. It is *locally trail-bounded* if for every vertex u_i there is a bound $b_i \in \mathcal{Z}^0$ such that the length of every (directed) trail emanating from u_i does not exceed b_i. Then g exists uniquely on every locally path-finite digraph, and for every $u \in V$, $g(u)$ does not exceed the length of the longest directed path from u. Similarly, γ exists uniquely on every locally trail-bounded digraph, and for every $u \in V^f$, $\gamma(u)$ does not exceed the length of the longest directed trail emanating from u.

These are existence results. Algorithms for g and γ are valid only for finite digraphs, since otherwise these algorithms do not necessarily terminate. The collapse of the algorithms for infinite digraphs has some curious ramifications. For example, an undecidable game, with path-finite game-graph, terminating in 3 moves, was constructed by Rabin (1957). Simpler examples were given later by Jones (1982). Note that the Fundamental Theorem (§2.7), specialized to path-finite digraphs, claims that precisely one of the two players can win. Is there a contradiction with the undecidability results? No, since the former is only an *existence* result; it does not claim to determine who of the two players has a winning move.

Here is an example of a different flavor, from Jones (1982). Given the polynomial

$$Q = x_1^2 + x_2^2 + 2x_1x_2 - x_3x_5 - 2x_3 - 2x_5 - 3,$$

players I and II alternate in assigning nonnegative integer values to the variables in the order $x_1, \ldots, x_5$. Player I wins if $Q = 0$; otherwise player II wins. This game has the interesting property that at present nobody knows who can win. This is so because

$$Q = (x_1 + x_2)^2 + 1 - (x_3 + 2)(x_5 + 2).$$

Since player II picks x_2, he has a winning strategy if and only if there are infinitely many primes of the form $n^2 + 1$, an old unsolved problem of number theory. (It does not really matter that x_4, selected by player II, does not appear in Q.)

5.7 Games in which the Loser Wins!

An example similar to the above but with still another flavor is this. Let

$$Q = x_1 - x_2x_4 - x_2 - x_4 - 1.$$

As before, the players alternately assign values to x_1, x_2, x_3, x_4 in this order. Player I has to select x_1 as a composite integer > 1, x_3 as any positive integer, and player II selects any positive integers. Player II wins if $Q = 0$; otherwise player I wins. It should be clear that player II can win, since $Q = x_1 - (x_2 + 1)(x_4 + 1)$ and x_1 is composite. But if player I picks x_1 as the product of 2 large primes of about the same size, then player II can realize his win in practice only if he can crack the RSA public-key cryptosystem (Rivest, Shamir and Adleman [**2**]). Thus in practice the loser wins! Also note that player II has an efficient probabilistic method — e.g. that of Solovay and Strassen [**3,4**] or Rabin (1976), [**1**] — to determine with arbitrarily small error that player I did not cheat, i.e., he selected a *composite* integer.

In the last 2 examples, the game-graph is also path-finite, and the difficulty is computing the next move. There are examples of games, even with finite game-graphs, where the winner's next move is trivial to compute, yet he can't realize his win in practice, because it takes an enormous amount of time. In such games both players lose in practice!

An example is provided by the class of epidemiography games (Fraenkel and Nešetřil (1985), Fraenkel, Loebl and Nešetřil (1988), Fraenkel and Lorberbom (1989)). This class of games has 2 interesting properties: (i) Robustness. The winner can play arbitrarily and still win, except that near the leaves of the game-tree the winner has to pay some attention. (ii) Length. The length of play grows explosively. For example in *Nimania*, played on a simple path of length n, the loser can delay the winner so that for $n = 3$ play lasts for 13 moves. For $n = 4$, the loser can delay the winner so that play lasts over 2^{44} moves!

In many other cases, we know next to nothing about the complexity classification of games with finite acyclic game-graphs. This topic is taken up in the next section.

6. Wonderland

In the realm of existential problems, such as sorting, TS, HC (§5.3), most problems, it seems, can be classified into tractable, conditionally intractable, and intractable ones. There are exceptions, to be sure, e.g. graph isomorphism and primality testing, whose complexity is still unknown. But these are few. It appears that, in contrast, there is a very large set of games whose complexities are hard to determine. The set of these games will be termed *Wonderland*, because we are wondering about the complexity classification of its members. We cannot hope to cover here even a small part of this huge set, and we confine ourselves to a subset of Wonderland, consisting of the games which are in Wonderland due to points (a) and (b) of §1.6. We mention only a few scattered examples of this large subset.

6.1. Succinct Games

Recall from §1.6 that these are games whose input size is rather small, so the game is tractable only if there is a strategy which is polynomial in this small input size.

An example is the set of *octal games* (Guy and Smith (1956), ONAG ch. 11, WW ch. 4, Guy (1989)), in which a portion of a simple path of length n (input size: $\Theta(\log n)$) may be removed according to rules encoded in octal. Nim with a single pile is the case where the removed part must lie strictly at one end of the path; in general, the rules specify removal of a certain portion within the path, which thereby breaks up into 2 parts. Thus the game becomes a sum of several sub-paths.

For the strategy to be polynomial, it suffices if the g-values follow some simple rule as a function of n, e.g., being periodic or increasing linearly as in Nim. Many octal games have in fact periodic g-values. Gangolli and Plambeck (1989) have recently produced large periods or large preperiods of some hitherto unchartered octal games, using a computer. But the character of many other members of this infinite family is still unknown. None has been proved intractable, though a generalization of "Kayles" from paths to arbitrary graphs has been proved Pspace-complete (Schaefer (1978, Theorem 3.2)). This generalization, however, is quite extensive; in particular, it increases the input size exponentially. See also problem 2 of "Unsolved Problems in Combinatorial Games" in this volume.

As another example consider *Modular Nim*, or *Kotzig's Nim* (WW ch. 15). A finite *move set* $M = \{a_1, \ldots, a_k\}$ is given, where the a_i are distinct positive integers, and a *modulus* $n \in \mathcal{Z}^+$. We imagine the numbers $0, 1, \ldots, n-1$ written on equally spaced points lying consecutively around a circle in clockwise direction, with 0 initially labeled. Player I begins by labeling a point a_i on the circle, provided $a_i \in M$. Then player II labels an unlabeled point on the circle at distance a_j from a_i, in clockwise direction, for some $a_j \in M$. At each stage a player labels an unlabeled point p at distance a from the point last labeled, in clockwise direction, for some $a \in M$. The player first unable to move loses and his opponent wins. Note that a player may lose either because the circle has just been tiled, or because it has been locally obstructed, though there are still unlabeled points on it.

Not much is known about this game. In particular, even for $k = 2$ and general move set $M = \{a, b\}$, it is not known whether the g-function, or even just the N, P-pattern is periodic (in n). For bounded k, the input size is succinct: $\Theta(\log n)$. See also problem 17.

6.2. Non-Disjointness

The game of Welter (1952, 1954) is Nim with distinct-size piles. It can be thought of as being played on a semi-infinite strip with squares numbered $0, 1, 2, \ldots$ and an arbitrary number m of coins distributed initially on distinct squares. A move consists of selecting a coin and moving it to any *unoccupied* lower square. Thus in Fig. 14 the coin on square 7 can be moved only to any of the squares 6, 5, 2 or 0. The game ends when the coins are jammed in the lowest squares (0, 1, 2, 3 in Fig. 14).

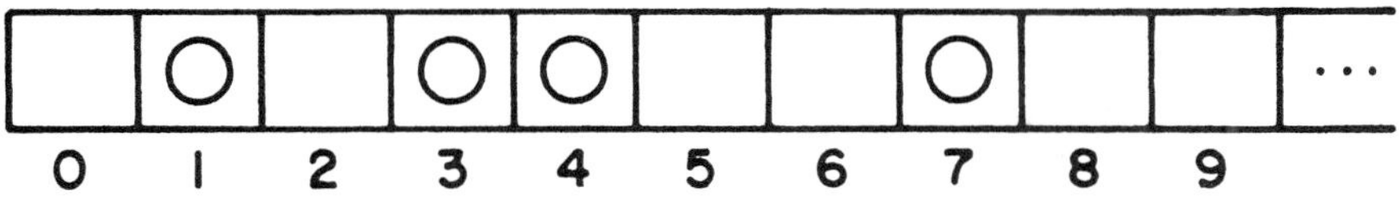

Figure 14. A game of Welter.

A full analysis of the g-values is given in ONAG ch. 13 (see also WW ch. 15); it is rather difficult. Berlekamp (1972) wrote about Welter: "Its solution is easy to state, difficult to prove, even harder to generalize to almost any other nonisomorphic game (no matter how small the difference between the other game and Welter's Nim), and virtually impossible to discover. Indeed, many of us who have studied it think that if Welter had not found the solution, it would still be unknown."

Since the number m of coins is not bounded, the input size is not succinct. But the game, due to the constraints on the moves, is not a strict sum of its components as Nim, which is the source of the difficulty of the game.

Until now we have considered only normal play. In *misère* play the outcome is reversed: the player first unable to move wins and his opponent loses. If G is the game-graph of any game, say acyclic, let G' denote the *pruned graph* of G, i.e., the digraph which remains after pruning all the leaves from G. Then normal play on G' is equivalent to misère play on G. The complexity of computing the N, P-pattern on G' is the same as that for G, so it appears that normal and misère play have the same complexity. In fact, this holds for Nim and a few other games.

In general, however, misère play is considerably harder than normal play. See e.g. Grundy and Smith (1956), ONAG ch. 12, WW ch. 13, Ferguson (1974), and Yamasaki (1980, 1981). The trouble, as often, stems from sums. If G is the sum of the component game-graphs $G_1, \ldots, G_m$, then it has size exponential in the size of $G_1, \ldots, G_m$. The way to overcome this obstacle for normal play was to observe that the Nim-sum σ of the g-functions of $G_1, \ldots, G_m$ is the g-function of G. This determines the N, P-labels for normal play. The difficulty of misère play seems to stem from the fact that if G' is the pruned graph of G, then G' is neither the sum of the pruned game-graphs $G'_1, \ldots, G'_m$, nor of the $G_1, \ldots, G_m$. Whereas a leaf of G is any position of the form $(u_1, \ldots, u_m)$ where the u_i are leaves of G_i for all i, $1 \le i \le m$, the leaves of G' have the form $(u'_1, \ldots, u'_m)$, where u'_j is a leaf of G'_j for some j, $1 \le j \le m$, and $u'_i = u_i$ for all $i \ne j$. Thus misère play is the sum of games which are not quite fixed.

6.3. Succinctness and Non-Disjointness

Poset games are games played on partially ordered sets. As an example, consider chomp (Gale (1974), Gardner (*Sc. American* Jan. 1973, Feb. 1973)), in which 2 players alternately move on a given $n \times m$ matrix of 1's (see Fig. 15). For a technical reason there is a single 0 at the origin. A player at his turn points to some 1, say at location (i, j), and removes the entire north-east sector (i.e., puts all the 1's to 0 inside the sector). The player removing the last 1 wins. The input size is $\log(mn)$, which is succinct. In addition, this game is not the sum of totally ordered sets, as Nim; rather it is the product of 2 Nim-piles.

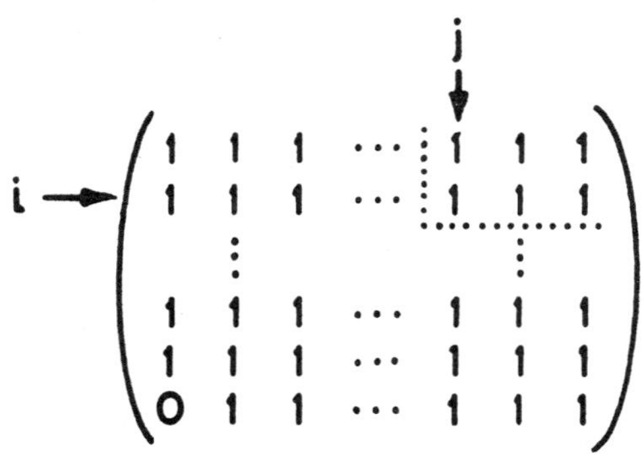

Figure 15. An initial position of chomp.

Neither polynomiality nor intractability are known for general m, n. However, it is easy to see that player I can win: If taking the element (m, n) (the "largest" element) is an opening winning move, then player I can make it and win. If it is a losing move, then there is a winning answer, say taking element (i, j). Then player I's first move is to take $(i, j) \ldots$ This argument holds in general for poset games with a largest element, but it is nonconstructive. Incidentally, computer simulations of chomp suggest that more often than not, an opening winning move is *not* to take the largest element. See also problem 27.

For the superset game, define $A_n^k = \{B \subseteq \{1, \ldots, n\}: 0 < |B| \leq k\}$. A player at his turn points at an as yet unremoved subset and removes it, together with all sets *containing* it. For normal play we then clearly have $A_n^0 \in P$ if and only if $n \equiv 0 \pmod 1$, $A_n^1 \in P$ if and only if $n \equiv 0 \pmod 2$. Gale and Neyman (1982) showed that $A_n^2 \in P$ if and only if $n \equiv 0 \pmod 3$, and therefore conjectured that $A_n^k \in P$ if and only if $n \equiv 0 \pmod{k+1}$.

Incidentally, note that $A_n^n \in N$ by the above nonconstructive argument, and so if the conjecture is true, then the unique winning move *is* to remove the largest element $\{1, \ldots, n\}$, much unlike the observed behavior of chomp. We mention that at the end of Fraenkel and Scheinerman [$\geq$ 1990], the g-values of the first few positions of A_n^k have been computed.

In von Neumann's Hackendot, played on a forest, a player points to an as yet unremoved vertex, and removes the unique path from that vertex to the root of the tree the vertex belongs to. This removal breaks up the tree into a forest, in general. The game is an N-position when begun on a tree, by the above nonconstructive argument. Úlehla (1980) gave an interesting polynomial strategy for the game. See also WW ch. 17, and problem 28.

6.4. Virtue in Weakness?

The above are but a few examples showing that even for some of the structurally simplest games we do not know whether they have a polynomial strategy or not. We saw that other games are intractable or conditionally intractable, having a polynomial strategy if and only if all kinds of other hard problems have a polynomial solution.

In order not to end this missive on a pessimistic note let's make a feeble attempt to turn ignorance into virtue, using some speculation. After all, games are supposed to be useful helping politicians to make decisions!

Whereas in all respectable walks of life we strive towards solutions or at least approximate solutions which are polynomial, there are 2 less respectable human activities in which high complexity is appreciated. These are cryptography (covert warfare) and games (overt warfare). The desirability of high complexity in cryptography — at least for the encryptor! — is clear. We speculate that it is also desirable, in a certain sense, for games.

There have been no international Nim-Olympiads, even prior to the turn of the century, when Bouton (1902) gave a polynomial strategy for Nim. For a game to be interesting to the public, intractability seems very desirable. It is not a sufficient condition: there are rather complex games on certain Boolean functions which do not appear interesting to play (Stockmeyer and Chandra (1979)). Nor is the condition even necessary: Gale's game Bridg-it has been shown to be polynomial by Oliver Gross (Gardner (Sc. Amer. Column, Oct. 1958, July 1961); reprinted in Gardner (1966)). Martin Gardner has informed us that it is the *only* polynomial game he knows which has enjoyed commercial success. However, we have seen in §5 that interesting board games such as chess and Go are intractable or conditionally intractable on $n \times n$ boards.

On the strength of this admittedly somewhat flimsy evidence, we venture to suggest that intractability of games is a virtue in the sense that it may bestow upon a game a depth, appeal and lustre usually lacking in polynomial games.

Acknowledgment. This survey is based, in part, on Fraenkel (1980).

References

1. M.O. Rabin [1980], Probabilistic algorithm for testing primality, *J. Number Theory* **12**, 128–138.

2. R.L. Rivest, A. Shamir and L. Adleman [1978], A method for obtaining digital signatures and public-key cryptosystems, *Comm. ACM* **21**, 120–126.

3. R. Solovay and V. Strassen [1977], A fast Monte-Carlo test for primality, *SIAM J. Comput.* **6**, 84–85.

4. R. Solovay and V. Strassen [1978], Erratum: A fast Monte-Carlo test for Primality, *SIAM J. Comput.* **7**, 118.

Proceedings of Symposia in Applied Mathematics
Volume **43**, 1991

. . . , Welter's Game, Sylver Coinage, Dots-and-Boxes, ...

Richard J. Nowakowski

1 Introduction

In order to play a game well, humans (and most computers) need patterns, extra rules or heuristics. Analyzing each position by examining all of its options and all the options of those options and all the options of the options of the . . . , can take a tremendous amount of time. This chapter provides 'hands-on' experience in discovering patterns. Readers are encouraged to try the questions written in **bold**. Hints or solutions are given at the end of the chapter. For more hints and solutions the reader should consult *On Numbers and Games*, *Winning Ways* and *Fair Game*.

We will consider (at least) three games.

Welter's Game is completely within the scope of the Sprague-Grundy theory, but to find the nim-value of a position and a good reply is quite difficult.

Sylver Coinage is a finite game (just!) as a theorem of J.J. Sylvester shows. Good opening moves are not difficult to find, but what about the third move?

Dots-and-Boxes is a game which does not seem to be covered by the Sprague-Grundy theory because it is neither a normal nor a misère game. But it can be related to Nimstring, a normal game.

The paper is in final form and no version of it will be submitted for publication elsewhere.
1980 *Mathematics Subject Classification* (1985 *Revision*). Primary 90D05.

But as a diversion, we first present

1.1 Brussels Sprouts

The game is played with a number of crosses with four arms. A move is to join two arms (of the same or different crosses) by a curve and to add a crossbar (with two new arms) at some point on the curve (Figure 1). The curve should not cross any other curve. Under normal play, the loser is the player who cannot move. This will happen when there is just one arm in each region so that no further curves can be drawn.

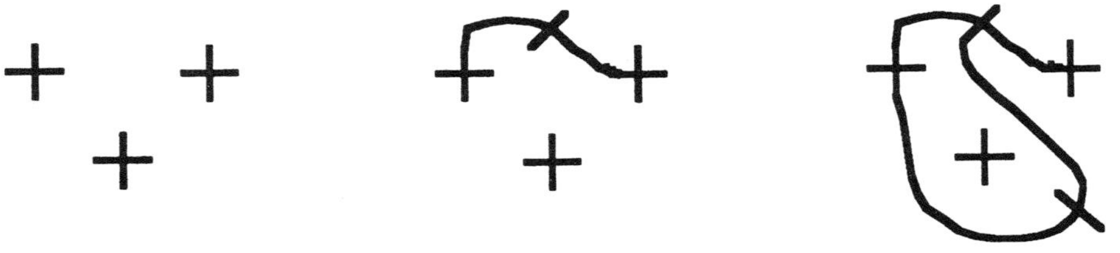

Figure 1: The first two moves in a game of three-cross Brussels Sprouts.

Find a winning strategy for both the normal and misère versions.

A hint is given at the end of this chapter.

1.2 Variants of Nim

There are several ways to disguise the game of Nim. But each disguise suggests different 'natural' variants. A game of Nim with k heaps could also be regarded as playing in k-dimensional space, on points whose coordinates are nonnegative integers. A move is to reduce any coordinate by an integer; the object of the game being to get to the origin in normal play or to force your opponent to the origin in misère play. A natural additional move is to allow diagonal moves. That is, in the original heap game, the same number may be taken from any number of heaps. **Wythoff's Game** is the 2-dimensional version of Nim with diagonal moves and has a well understood and interesting theory. Would any one care to play the 3-dimensional version? Another variant is to take a semi-infinite strip of squares which are numbered with the non-negative integers $0, 1, 2, 3, \ldots$ from the left. A game of k-heap Nim can now be represented by k coins with each coin on a square which indicates the size of the corresponding heap. A move takes a coin to a lower numbered square. This game allows more than one coin to a square. A natural variant is to allow only one coin per square.

First, an easy game (once you know the trick). Coins are not allowed to pass one another, so we will call it the **Sliding Game.** This is a cleverly disguised form of Nim. The sizes of the heaps are the lengths of alternate gaps starting at the right most coin. If there is an odd number of coins then the last heap is the number of the square of the left most coin (i.e. the 'gap' including zero). For example, the game in Figure 2 corresponds to the 4-heap Nim game (reading from the right) 2, 1, 2, 2.

0	1	○	3	○	5	6	○	○	9	○	11	○	13	14	○	16	17	...

Figure 2: The Position (2,4,7,8,10,12,15) in the Sliding Game.

A move may decrease or increase a gap so that a heap size may *increase* as well as decrease. Moving the coin from 4 to 3 increases the third gap from 2 to 3. However, the next player can reduce this gap back to 2 (by moving 7 to 6) without affecting the other heaps (a gap of zero has been increased to one, but this gap does not correspond to a heap). Any heap-increasing move is therefore a **reversible** move and the size of a heap can be increased only a finite number of times.

The position in Figure 2 is an $\mathcal{N}$-position. **Can you find the three winning moves?**

Suppose we play the **Infinite-Sliding Game**, that is, play with a finite number of coins on a finite number of semi-infinite strips of squares where the strips are also labelled 0, 1, ... Again, two coins cannot occupy the same square and a move is to slide a coin *leftwards* to an unoccupied square without jumping over the next coin or if there is no coin to the left on that strip, it may continue sliding down the squares in the next lower-numbered strips (each time starting at 'infinity') and may stop at any square short of the next coin. The infinite-sliding game is not Hilbert Nim (Nim played with arbitrary ordinals) in disguise. Imagine two coins, one on strip 0 on square 3 the other on strip 1 on square 0 — the infinite-sliding game seems to need heaps with sizes like $\omega - 2$. If the 'nim-in-disguise' analysis does not carry over, how do we find a good move? The following analysis is due to Hugh Thomas.

Add another strip labeled ω which will account for the 'infinities' involved. Suppose that the strip i has n_i many coins. The ω strip has coins in the first n_0 positions; then an unoccupied square; then n_1 consecutive coins; then an unoccupied square; etc. A move in the game proper is made in the obvious way on the new game board. The nimvalue of the game is calculated as if it were the disjunctive compound of all the strips (each play the sliding game) except if the ω strip has nimvalue i in the sliding game then it has the nimvalue ωi in the infinite-sliding game.

The proof that this works is not too difficult to see. If the nimsum is 0 then the next player either moves a coin on one strip and changes the nimvalue of that strip in the 'disjunctive compound', or moves a coin from one strip to another which changes the nimvalue of the ω strip. If the nimsum is not 0 but is finite then the next player just plays as if he is playing in the disjunctive compound (for which there is a move to a position with nimsum 0). If the nimvalue of the ω strip is not 0 then the next player moves the coin on the ω strip that makes the strip's nimvalue 0. The corresponding coin in the real game is then moved to the corresponding strip. Since the coin is sliding in from 'infinity', it can create a nimheap of any size between it and the next coin or the end of the strip. Consequently, a square can be chosen that makes the nimsum of the disjunctive compound 0.

Figure 3 shows a game with five coins on four strips. The first and second coins are on strip 0 on the squares 1 and 4 respectively. The third and fourth coins are on strip 1 on squares 0 and 2 and the fifth coin is on strip 3 on the square 1. The ω strip has coins on the squares 0, 1, 3, 4 and 7. The nimvalues of the strips 0 through 3 are 2, 1, 0, 1 respectively and the ω strip has the nimvalue $\omega 3$. The nimvalue of the game is $2 \overset{*}{+} 1 \overset{*}{+} 0 \overset{*}{+} 1 \overset{*}{+} \omega 3 = 2 \overset{*}{+} \omega 3$. Since the nimvalue is infinite, any good move has to involve moving a coin from one strip to another. Here, the only good move is to move the fifth coin to square 2 on strip 2.

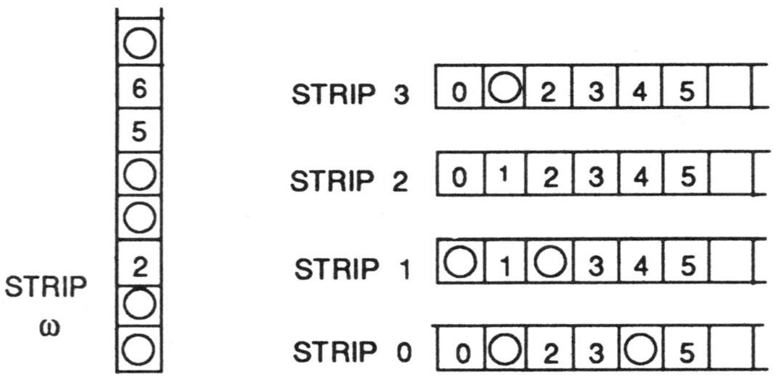

Figure 3: A Position in the Infinite-Sliding Game.

Can you extend this analysis to the Sliding Game with infinitely many strips?

2 Welter's Game

Welter's Game is also played on a semi-infinite strip with a finite number of coins and at most one per square. The legal move (for either player) is to move any one coin from its present square to *any* unoccupied square with a lower number, that is coins may jump past each other. As in the Sliding Game, Welter's game ends when one player (the loser) is unable to move because the coins are jammed in the lowest possible numbered squares $0, 1, 2, ..., k-1$.

We shall write $[a|b|c|...]_k$ for the nim-value of the position in this game when there are k coins on the squares numbered $a, b, c, ...$ It is easy to check that $[a] = a$. To see that $[a|b] = (a \overset{*}{+} b) - 1$ consider a and b as binary numbers; then it is easy to see that the values of the options are the nim-sums

$$a \overset{*}{+} b', \quad 0 \leq b' < b, \quad a \neq b' \quad \text{and} \quad a' \overset{*}{+} b, \quad 0 \leq a' < a, \quad a' \neq b$$

which take on all values from 1 to $(a \overset{*}{+} b) - 1$ but not $a \overset{*}{+} b$. Therefore,

$$[a|b] = \text{mex}\{1-1, 2-1, \ldots, ((a \overset{*}{+} b) - 1) - 1\} = (a \overset{*}{+} b) - 1.$$

Both nim-addition and ordinary addition are involved in the theory of Welter's game. The full theory is surprisingly complex, and we will have to refer the reader to *On Numbers and Games* for the proof of the general result. Here, we only give a computation rule for $[a|b|c|...]_k$.

2.1 The Mating Method.

We take as an example the position given in Figure 2, but played as Welter's Game, i.e we wish to find $[2|4|6|8|10|12|15]$. We examine the numbers to see which pair is congruent modulo the highest possible power of 2, and then we take any such pair out as **mates**. In our example, we have the congruences $4 \equiv 12$ and $2 \equiv 10 \bmod 8$ but no congruences modulo 16, so we can take either (4,12) or (2,10) as our best mated pair.

Having removed the best mated pair, we treat the remaining numbers in the same way, obtaining a succession of pairs of mates, with at most one number, **the spinster**, left unmated. In our example, the mates are (4,12), (2,10), (6,8), and there is a spinster 15. Then if (a, b), (c, d),... are the mates, we have the equation

$$[a|b|c|d|...] = [a|b] \overset{*}{+} [c|d] \overset{*}{+} ...$$

if there is no spinster, and

$$[a|b|c|d|...] = [a|b] \overset{*}{+} [c|d] \overset{*}{+} ... \overset{*}{+} [s]$$

if there is a spinster s. The value is then computed using the formulas

$$[a|b] = (a \overset{*}{+} b) - 1, \ [s] = s,$$

as noted previously.

So we have the number

$$[4|12] \overset{*}{+} [2|10] \overset{*}{+} [6|8] \overset{*}{+} [15] = 7 \overset{*}{+} 7 \overset{*}{+} 13 \overset{*}{+} 15 = 2$$

for the nim-value of our example. This rule finds the nim-value: but how do we find a good move, which takes the nim-value to zero? We need a method that inverts the Welter function. We shall postpone giving this until later. In the example, the only good move is from 15 to 13.

Find the nim-value of the position (2, 4, 7, 8, 10, 13, 14, 16).

2.2 Some simple strategies for easy games.

One and two coin games are easy to handle with the information already given.

A Nim-like strategy also suffices for Welter's Game with four coins. In fact,

$$[a|b|c|d] = 0 \quad \text{just if} \quad a \overset{*}{+} b \overset{*}{+} c \overset{*}{+} d = 0.$$

Welter's theory explains this by noting that if a, b and c, d are the mates, these equalities reduce to

$$[a|b] = [c|d] \quad \text{and} \quad a \overset{*}{+} b = c \overset{*}{+} d$$

which are equivalent statements since $[x|y] = (x \overset{*}{+} y) - 1$.

If one of the four coins is on 0 then this is three-coin Welter with the others, but shifted one place. Or

$$[0|a|b|c] = [a-1|b-1|c-1]$$

or

$$[a|b|c] = [0|a+1|b+1|c+1].$$

Thus the Welter position with coins on 1, 3, 7 is equivalent to the Welter or Nim position with coins on 0, 2, 4, 8. The winning move is 8 to 6, so in the three-coin position we should move 7 to 5.

What if we have more than four coins? Let us look at the position (3, 5, 8, 13, 21). We can calculate that

$$[3|5|8|13|21] = [5|21] \overset{*}{+} [3|13] \overset{*}{+} [8] = 15 \overset{*}{+} 13 \overset{*}{+} 8 = 10.$$

So there must be a good move, what is it? Welter, O'Beirne and Berlekamp all gave methods to find good moves, but all were laborious. In this example, which coin should we move? Perhaps the coin on 21 has the most freedom, so let's try to solve the equation

$$[3|5|8|13|x] = 0.$$

What is x mated with? Suppose it is the spinster. This gives

$$[5|13] \overset{*}{+} [3|8] \overset{*}{+} x = 7 \overset{*}{+} 10 \overset{*}{+} x = 0$$

and so $x = 13$. This is **very** well mated with 13, so we next suppose that

$$[x|13] \overset{*}{+} [3|5] \overset{*}{+} 8 = [x|13] \overset{*}{+} 5 \overset{*}{+} 8 = [x|13] \overset{*}{+} 13 = 0,$$

and therefore $[x|13] = 13$, $x \overset{*}{+} 13 = 14$, $x = 3$. This is now very well mated with 3 so we try

$$[x|3] \overset{*}{+} [5|13] \overset{*}{+} 8 = [x|3] \overset{*}{+} 7 \overset{*}{+} 8 = [x|3] \overset{*}{+} 15 = 0$$

and therefore $[x|3] = 15$, $x \overset{*}{+} 3 = 16$, $x = 19$. Finally, 19 is well mated with 3 and

$$[3|19] \overset{*}{+} [5|13] \overset{*}{+} 8 = 15 \overset{*}{+} 7 \overset{*}{+} 8 = 0$$

confirming our guess that there was a good move from 21, namely to 19. If there had not been a good move from 21 we would have to repeat the calculation with 13 then 8 and so on until a good move was found. There has to be one!

Fortunately, Conway discovered that you could replace addition or multiplication in frieze patterns by nim-addition and that this not only leads to a surprising new way of calculating the Welter function but also to a comparatively painless way of inverting it!

2.3 Frieze Patterns

Patterns of numbers such as

1		1		1		1		1		1		1		1			
	1		2		2		3		1		2		4		1		
		1		3		5		2		1		7		3		1	
			1		7		3		1		3		5		2		1
		1		2		4		1		2		2		3		1	
	1		1		1		1		1		1		1		1		

in which each diamond of numbers

$$\begin{array}{ccc} & b & \\ a & & d \\ & c & \end{array} \quad \text{satisfies} \quad ad = bc + 1 \quad \text{so that} \quad d = (bc+1)/a,$$

have many wonderful properties. For example, if you start with two horizontal rows of 1's connected by any zigzag of intermediate 1's, say

$$\begin{array}{ccccccccccccccccccccc}
1&&1&&1&&1&&1&&1&&1&&1&&1&&&&\\
&1&&?&&?&&?&&?&&!&&?&&?&&!&&&\\
1&&?&&?&&?&&?&&!&&?&&?&&!&&&&\\
&1&&?&&?&&?&&!&&?&&?&&?&&!&&&\\
&&1&&?&&?&&!&&?&&?&&?&&?&&!&&\\
&&&1&&?&&?&&!&&?&&?&&?&&?&&!&\\
&&&&1&&1&&1&&1&&1&&1&&1&&1&&1
\end{array}$$

you'll find that all the entries are whole numbers and that each ! is 1 so that the pattern repeats itself alternately, first one way up then the other.

G.C. Shephard has observed that we can replace multiplication by addition, provided we start with a row of 0's instead of 1's, making each diamond

$$\begin{array}{ccc} & b & \\ a & & d \\ & c & \end{array} \quad \text{satisfy} \quad a + d = b + c + 1 \quad \text{so that} \quad d = b + c + 1 - a.$$

Try some examples for yourself.

Now remember that Conway discovered that if you use **nim-addition** and start with the Welter position below a row of zeros, so that each diamond

$$\begin{array}{ccc} & b & \\ a & & d \\ & c & \end{array} \quad \text{satisfies} \quad (a \overset{*}{+} d) = (b \overset{*}{+} c) + 1 \quad \text{i.e.,} \quad c = ((a \overset{*}{+} d) - 1) \overset{*}{+} b,$$

then you'll find the value of the Welter position at the bottom of the triangle! E.g. the position $\{2, 3, 5, 7, 11, 13, 17, 19\}$ has nim-value 17

$$\begin{array}{ccccccccccccccccc}
0&&0&&0&&0&&0&&0&&0&&0&&0\\
&2&&3&&5&&7&&11&&13&&17&&19&\\
&&0&&5&&1&&11&&5&&27&&1&&\\
&&&7&&6&&14&&6&&16&&8&&&\\
&&&&5&&6&&12&&16&&12&&&&\\
&&&&&4&&7&&29&&11&&&&&\\
&&&&&&4&&21&&5&&&&&&\\
&&&&&&&23&&18&&&&&&&\\
&&&&&&&&17&&&&&&&&
\end{array}$$

This rule is equivalent to the identity

$$[a|b|...|y|z]_{k+1} = [[a|b...|y]_k|[b|...|y|z]_k] \overset{*}{+} [b|...|y]_{k-1}$$

which you can find in *On Numbers and Games* (p. 159)

Although this calculation seems tedious, it requires little thought and can be easily implemented on a computer. Moreover, it produces a way to invert the Welter function.

Suppose that you've evaluated

$$[a|b|c|...] = n$$

and have in mind a number $n' \neq n$. Then there are unique numbers

$$a' \neq a, b' \neq b, c' \neq c, ...$$

for which

$$[a'|b|c...] = n',$$
$$[a|b'|c...] = n',$$
$$[a|b|c'|...] = n',$$

Moreover it can be shown that the equation

$$[a|b|c|...] = n$$

remains true if an **even** number of the letters $a, b, c, ..., n$ are replaced by the corresponding primed letters. This can be expressed as a single equation

$$\left[\begin{array}{c|c|c|c} a & b & c & \cdots \\ a' & b' & c' & \cdots \end{array}\right] = \begin{array}{c} n \\ n' \end{array}$$

Using this **Even Alteration Theorem** and the properties of frieze patterns you can invert the Welter function.

For example, we had the position and its corresponding Welter function

$$[3|5|8|13|21] = 10$$

and we want to find numbers a', b', c', d', e' respectively to replace 3, 5, 8, 13, 21 so that the new Welter function is zero in each case. Conway's idea is to recalculate the Welter function using a frieze pattern which will give 10 in the bottom row which he extends to 10, 0, 10, 0, 10, 0, ... then works back upwards. The good moves are to be found in the extension of row containing the original Welter position!

0	0	0	0	0	0	0	0	0	0	0	0	0	0
3	5	8	13	21	37	35	2	11	19	3	5	8	
5	12	4	23	47	5	32	8	23	15	5	12	4	
13	15	31	34	12	7	37	21	4	10	13	15		
13	11	43	2	15	1	39	7	2	3	13	11		
10	0	10	0	10	0	10	0	10	0	10			

In other words,

$$\left[\begin{array}{c|c|c|c|c} 3 & 5 & 8 & 13 & 21 \\ 37 & 35 & 2 & 11 & 19 \end{array}\right] = \begin{array}{c} 10 \\ 0 \end{array}$$

and the good moves are from 8 to 2, 13 to 11 and the move 21 to 19 that we already found.

2.4 Congruence Heuristics.

Although the Mating Method makes it very easy to work out nim-values, it's not so easy to find which move you should make to restore the nim-value to 0. But if the number of coins is a multiple of 4, there is a remarkable connexion with Nim: $[a|b|c...]_{4k} \equiv 0 \bmod 16$ exactly when $a \overset{*}{+} b \overset{*}{+} c + \ldots \equiv 0 \bmod 16$. This ensures in particular that when the $4k$ coins are among the first 16 places, the Welter's Game $\mathcal{P}$-positions are exactly those $\mathcal{P}$-positions in Nim that have distinct numbers. The proof for the result is given modulo 8.

Let $[a|b|c|\ldots]_{4k}$ be a $\mathcal{P}$-position. If a, b and c, d are mated pairs then they cannot both be mated modulo 1. For, suppose that their binary representations end in $a = \ldots 0$, $b = \ldots 1$; $c = \ldots 0$, $d = \ldots 1$, then a, c and b, d would be the mates instead. Also, if there were just one pair, a, b, mated modulo 1 then $[a|b]$ ends in 0. All the values of the other mated pairs would end in 1 and there are an odd number of them. Hence this would not be a $\mathcal{P}$-position. Let w, x, y, z, respectively, represent the number of mated pairs whose nim-sums are congruent to 110, 100, 010, 000 modulo 8 and whose 'Welter' values are 101, 011, 001, 111 modulo 8. Since $[a|b|c|\ldots]_{4k}$ is a $\mathcal{P}$-position, then by considering the fours column, we see that $w+z$ is even; from the twos column, we see that $x+z$ is even; and from the ones column, we see that $w+x+y+z$ is even. Adding the first two we see that $w+x$ is even and adding the second and third we find that $w+y$ is even. Now, the nim-sum $a \overset{*}{+} b \overset{*}{+} c \overset{*}{+} \ldots$ can be obtained by first nim-summing the mates. Then, there are an even number of fours

($w+x$ is even) and the number of twos is even ($w+y$ is even). There are no ones so $a \overset{*}{+} b \overset{*}{+} c \overset{*}{+} \ldots \equiv 0, \mathrm{mod} 8$.

The same result is true modulo 16.

Find a proof for the result modulo 16. Is there a similar result modulo 32?

Using the above hints, what are the good moves from (1, 3, 4, 6, 7, 9, 10, 15)?

These numbers	1,	3,	4,	**6,**	7,	9,	10,	15	nim-sum to 11,
and we get	10	8	15	**13**	12	2	1	4	after nim-adding 11 to them.
	x	x	x	x	x		y	y	

The x's and y's indicate illegal moves; x indicates a move which places a coin on a higher numbered square, and y indicates a move which places a coin on an already occupied square. The only legal winning move is from 9 to 2.

What are the mates in (1,3,4,6,7,9,10,15)? and what are the mates after you have made the winning move?

What, if any, is a good move in (2,4,6,8,10,12,14,16)?

What happens when the number of coins isn't a multiple of 4? If there are 6 coins, say, on positions $1, 3, 4, 5, 8, 13$ this is equivalent to having 8 coins on places

$$-2, -1, 1, 3, 4, 5, 8, 13$$

on a strip numbered starting from -2. Renumbering from 0 we see the position

	0	1	3	5	6	7	10	15	nim-sums to 3,
yielding	3	2	0	6	5	4	9	12	after nim-adding 3 to each
	x	x	y	x	y				

so this time there are three good moves, from 7 to 4, 10 to 9, 15 to 12, in the new notation, or 5 to 2, 8 to 7, 13 to 10, in the original 6-coin position.

If there are 5 coins, say on $(2, 3, 5, 7, 11)$, we increase by 3 to get

	0	1,	2,	5,	6,	8	10,	14	which nim-sum to 12,
yielding				9	10	4	6	2	on nim-adding 12.
				x	x		y	y	

After ignoring the illegal moves the only good move, 8 to 4, translates back to the original game as 5 to 1.

Do you want to play first in the position $\{\omega, \omega+1\}$?

And what about $\{\omega, \omega+1, \omega+2\}$?

What is the theory for the Infinite-Welter's Game?

3 Sylver Coinage

In this game the two players take turns in naming positive integers subject to the rule that the number they name cannot be a positive linear combination of previously named numbers. So, for example, if 3 and 4 have been named, only 1, 2 and 5 are legal moves since neither player can name

$$3,\ 4,\ 3+3=6,\ 3+4=7,\ 4+4=8,\ 3+3+3=9,\ \ldots.$$

The loser is the first player to name 1. Sylver Coinage is a *misère* game. (The normal play version is clearly uninteresting!)

The game is surprisingly difficult. The difficulty rests on the fact that a game can last for a very long time. For example, here are possible moves in a game lasting 100 moves:

$$100,\ 99,\ 98,\ 97,\ \ldots,\ 3,\ 2,\ 1.$$

Since any number could replace 100 the length of the game is unbounded. But the situation is much worse, because there is no fixed number k such that after k moves the number of moves remaining in the game is bounded. Again, if $k = 100$ then let the first k moves be $2^{100}, 2^{99}, 2^{98}, \ldots, 2^3, 2^2, 2^1$, then the next move can be any odd number with the players taking decreasing consecutive odd numbers. Consequently, finding a good move may mean analyzing infinitely many possibilities.

Despite this argument, no game can last forever. The theorem that proves this is due to J.J. Sylvester, hence the name *Sylver* Coinage. Essentially, if a and b, $a < b$, have greatest common divisor (g.c.d.) g then $ab - a - b$ is the largest multiple of g which is not the positive sum of a's and b's. Consequently, if g is the g.c.d. of the first n moves then there are only finitely many multiples of g which are not a positive sum of these numbers. After they have been chosen or otherwise eliminated, some number will have to be chosen that reduces the g.c.d. Therefore, after finitely many moves, we reach a position in which $g = 1$ and now the game is bounded.

What is a good reply to 4?

3.1 Strategies.

While it is difficult to find winning strategies in general, some positions are easy.

If you name 1 you lose.

The numbers 2 and 3 are good replies to each other, for after both have been named then all larger numbers, $4 = 2+2$, $5 = 2+3$, $6 = 2+2+2$, ..., are excluded leaving the next player to name 1 and lose. At this point the situation becomes more difficult.

Did you find a good reply to 4? Clearly neither 2 nor 3 will do (and certainly not 1). Let's try 5. This gives a g.c.d. of 1 and so leaves only finitely many numbers. Which numbers are left can be discovered by arranging them modulo 4 as in Table 1. The boxes are excluded because they are multiples of 5. Numbers below boxes are excluded because each is an appropriate multiple of 5 plus a multiple of 4. From the table we see that only 1, 2, 3, 6, 7 and 11 are left.

$\boxed{0}$	1	2	3
	$\boxed{5}$	6	7
		$\boxed{10}$	11
			$\boxed{15}$

Table 1. The legal moves left in the game $\{4, 5\}$

If you name 1, you lose immediately. If you name either 2 or 3, your opponent names the other leaving you to name 1 (and lose). If you name 6 or 7 then your opponent names the other. Both eliminate 11 and you are left naming 2 or 3 (and losing). Therefore the unique winning move is 11. Consequently $\{4, 5\}$ is an $\mathcal{N}$-position and $\{4, 5, 11\}$ is a $\mathcal{P}$-position.

This technique of **pairing** moves works. In the previous example 2 and 3 are paired and so are 6 and 7.

Use the pairing strategy to find the unique good reply to $\{5, 6\}$.

It would appear that this strategy works because there are only a few legal moves left. However, a table can help to suggest pairings even if there are infinitely many entries. For example, consider the position $\{4, 6\}$. We already know that 2 and 3 are good replies to each other. Table 2 suggests that $(5, 7), (9, 11), ..., (4k + 1, 4k + 3), \ldots$ should be paired.

$\boxed{0}$	1	2	3
	5	$\boxed{6}$	7
	9		11
	13		15
	$\vdots$		$\vdots$

Table 2. The legal moves left in the game $\{4, 6\}$.

Indeed, the first player has to mention one of these numbers and the second player can name its mate. Therefore $\{4,6\}$ is a $\mathcal{P}$-position. Consequently, you should never open with 4 or 6 since they are good replies to one another.

The pairing scheme also gives a strategy for $\{8,12\}$. The pairs are

$$(2,3),(5,7),(9,11),\ldots,(4k+1,4k+3),\ldots$$

and

$$(4,6),(10,14),(18,22),\ldots,(8k+2,8k+6),\ldots;$$

and for $\{6,9\}$ the pairs are

$$(4,11),(5,8),(7,10) \quad \text{and} \quad (3k+1,3k+2) \quad \text{for} \quad k \geq 4.$$

But beware that after naming some of the smaller numbers, the pairings will change. After (4, 11) pair (5, 7); after (5, 8) pair (4, 7); and after (7, 10) pair (4, 5) and (8, 11).

Note that this means that 1, 2, 3, 4, 6, 8, 9, 12 are all BAD opening moves.

Higher levels of pairings are possible. A **clique** is a set of numbers which have the property that any good reply to a clique number must also be a clique number and together they must eliminate all higher numbers outside of the clique and no others in the clique. In $\{4,5\}$, 1 was a (special) clique; (2,3) and (6,7) formed two other cliques. In $\{4,6\}$, again (2,3) was a clique as were $(4k+1,4k+3)$ for all $k \geq 1$.

$\boxed{0}$	1	2	3	4	5
	$\boxed{7}$	8	9	10	11
		$\boxed{14}$	15	16	17
			$\boxed{21}$	22	23
				$\boxed{28}$	29
					$\boxed{35}$

Table 3. The game $\{6,7\}$ can be analysed using cliques.

We know that 2 and 3 are always good replies to one another; and 4 and 5 exclude all larger numbers in Table 3 and therefore also form a clique. Whoever first names one of 4 or 5 will lose when the opponent names the other. Now 8, 9, 10 and 11 form another clique and are paired as (8, 10) and (9, 11). Each pair of moves excludes all higher numbers outside the clique but leaves the other pair. If you name any of 2, 3, 4, 5, 8, 9, 10, 11 your opponent will name the corresponding paired number and you will have to 'open' another pair and finally be forced to name 1.

From the discussion so far, we know that if there is a good reply to one of the six remaining numbers 15, 16, 17, 22, 23, 29 then it must be another of them. Note that (15,17) do not form a clique since after being named they would leave 16. But 15 and 23 are good replies to each other since they leave the pair (16, 17); and so are 17 and 22, since the pair (15, 16) is left; while naming 16 leaves the pair (15,17). Now 15, 16, 17, 22, and 23 form a clique since 29 is excluded by each of them. Consequently, 16 is the unique good reply to $\{6, 7\}$.

3.2 Some good opening moves.

R.L. Hutchings has determined the status of many opening moves.

If a and b are coprime and $\{a, b\} \neq \{2, 3\}$ then $\{a, b\}$ is an $\mathcal{N}$-position.

The proof is by strategy-stealing. Let t be the largest number not excluded by $\{a, b\}$. If t is **not** a good reply to $\{a, b\}$ then some s **is** a good reply to $\{a, b, t\}$. Since s excludes t, the positions $\{a, b, s, t\}$ and $\{a, b, s\}$ are equivalent so that if $\{a, b, s, t\}$ is a $\mathcal{P}$-position then $\{a, b, s\}$ is also. Consequently the first player should **steal** the second player's strategy and name s instead of t.

$\boxed{0}$	1		2	3	4
	$6 = s$		$\boxed{7}$	8	9
	11	↓		13	$\boxed{14}$
	16	↓ ↘		18	
	$\boxed{21}$			→ $23 = 6 + 2a + b$	
				$\boxed{28}$	

Table 4. Hutchings's Theorem with $a = 5,\ b = 7$.

This proof is not constructive. Note that any move will exclude the last included or *topmost* number t. To see this, assume $a < b$ and write out the table of numbers in a columns. Then, in each column the *first excluded* number is a multiple of b, so that the *last included* numbers in each column differ by multiples of b. From any legal reply to $\{a, b\}$ we can get to the last legal number in that column by adding a multiple of a, and then we get to t by adding multiples of b. There are two corollaries to this result.

The P Theorem:

If $p \geq 5$ is a prime number, $\{p\}$ is a $\mathcal{P}$-position.

which follows since the second player must name a number coprime to p.

The N Theorem:

If n is a composite number not of the form $2^a 3^b$ then $\{n\}$ is an $\mathcal{N}$-position.

This result follows since the next player can name a prime number (greater than 3) which divides n and thus excludes n and the situation is now covered by the P Theorem. Naming a prime divisor of 2^a3^b would, of course, *lose* the game.

3.3 Another Strategy.

Let us call $\{a, b\}$, where $\gcd(a, b) = 1$, an **end-position** (because, if the topmost number t is greater than 1, it is an $\mathcal{N}$-position). Call $\{a, b, c, \ldots\}$ a **quiet end-position** if, for every legal move s, the topmost number t can be made up using just *one* copy of s. These will be end-positions and $\mathcal{N}$-positions because the topmost number is eliminated by every move and so the same argument as for Hutchings's theorem (strategy-stealing) still applies.

The Quiet End Theorem. *If a is coprime to each of b and b_1, then $S = \{a, bc, bd, be, ...\}$ is a quiet end-position if and only if $S_1 = \{a, b_1c, b_1d, b_1e, ...\}$ is.*

Again, write the numbers in a columns, boxing the first excluded numbers in each column. For S, if n is boxed then n is excluded by S but $n - a$ is *not* so excluded. Therefore $n = ak + bm$ where m is a sum of multiples of $c, d, e, \ldots.$ If k were positive then $n - a = a(k-1) + bm$ would also be excluded by S, hence $k = 0$ and $n = bm$.

Now we claim that bm is boxed just if b_1m is boxed. Certainly b_1m is excluded by S_1. It would not be boxed if $b_1m - a$ were excluded. If that were the case then $b_1m - a = ak + b_1'$, again m' is a sum of multiples of $c, d, e, \ldots.$ Therefore, $b_1m = a(k+1) + b_1m'$ and so b_1 divides $k + 1$ since it is coprime to a. Dividing by b_1 and multiplying by b we obtain $bm = ak' + bm'$ for some positive number k' showing that $bm - a = a(k'-1) + bm'$ was excluded and that bm was not boxed for S.

Therefore, for any two last included numbers $r = bm - a$ and $s = bn - a$ in S we have the relationship $r = s + bq$, and the corresponding numbers $r_1 = b_1m - a$ and $s_1 = b_1n - a$ in S_1 have the same arithmetic relationship, i.e. $r_1 = s_1 + b_1q$.

This can be used as a reduction technique. For example, by taking $b_1 = 1$, we see that no odd number is a good reply to $\{8, 12\}$. Clearly 1 is not and for any other odd number a, $\{a, 2, 3\}$ is the same as $\{2, 3\}$. By the quiet end theorem $\{a, 8, 12\}$ is a quiet end-position and so an $\mathcal{N}$-position.

3.4 Some Good Replies.

Good replies are known to many 'small' positions. All the entries, bar one, for the table given on page 596 of volume 2 of *Winning Ways* are now known and is our Table 5. The exceptional entry is the last entry which contains all primes greater than 3 and may include some numbers 2^a3^b.

	{7,9,11}	{7,9}	{7,11}	{7}	{9,11}	{9}	{11}	{}
{6,8,10}	[ ]	[11]	[9]	[ ]	[4,5,7]	[5]	[ ]	[4,7,11]
{6,8}	[10]	[ ]	[ ]	[9,10,11]	[4,5]	[5,7]	[7,10]	[4]
{6,10}	[8]	[ ]	[15]	[8,9]	[4]	[7]	[8,13]	[4]
{6}	[ ]	[8,10,11]	[8,9]	[16]	[4,7]	[ ]	[26]	[4,9]
{8,10,12}	[4,5,6]	[4]	[13]	[5,6]	[13,14,15]	[23]	[6]	[14]
{8,12}	[5]	[6]	[6]	[5]	[ ]	[11,15]	[9,13]	[ ]
{10,12}	[4]	[4,6]	[16]	[ ]	[ ]	[11,13,14]	[9,14,15,17]	[7,18]
{12}	[6]	[15]	[27]	[10]	[8,10]	[6]	[49]	[8]
{8,10}	[4,5,6]	[4]	[ ]	[5,6,11]	[23]	[13,15]	[6,7]	[22]
{8}	[5]	[6]	[6,10]	[5]	[12]	[21]	[23]	[12,14]
{10}	[4]	[4,6]	[8]	[12]	[12]	[16]	[24,28,47]	[5,14]
{}	[6]	[19,24]	[24,34]	[ ]	[13,30]	[6]	[ ]	[5,7,11,13,...

Table 5. Status of Non Empty Subsets of {6,7,8,9,10,11,12} and all Good Replies

Even members of the set are on the left; odd members are at the top. Closed brackets indicate that all good replies are known; [ ] indicates a $\mathcal{P}$-position.

R.L. Hutchings and F. Voelkle have found the good replies to a first move of a composite number between 4 and 25. These are given in Table 6 (see R.K. Guy, Unsolved Problems, *Amer. Math. Monthly*, **94**(1987) 963).

Moves	Replies	Moves	Replies	Moves	Replies
{4}	[6]	{12}	[8,($4n+2 \geq 30, 3n \geq 27$)	{20}	[5,($2n \geq 16$)
{6}	[4,9]	{14}	[7,8,10(26)	{21}	[7,($3n \geq 15$)
{8}	[12,14]	{15}	[5,$3n \geq 21$)	{22}	[11,($2n \geq 18$)
{9}	[6]	{16}	[(20,$2n \geq 26$)	{24}	[($3n \geq 15, 2n \geq 20$)
{10}	[5,14,(26,32,36,46)	{18}	[($2n \geq 20, 3n \geq 30$)	{25}	[5]

Table 6. Some Good Second Moves.
The known good replies are followed, in parentheses, by the only other candidates.

In the cases where there have been two moves played, one between 6 and 14, the other between 6 and 25, G.L. Sicherman has extended the known good replies. These are given in Table 7.

	6	6	8	9	10	11	12	13	14
7	[16]								
8	[4]	[5]							
9	[ ]	[19,24]	[21]						
10	[4]	[12]	[22]	[16]					
11	[26]	[24,34]	[23]	[13,30]	[24,28,47]				
12	[4,9]	[10]	[ ]	[6]	[7,18]	[49]			
13	[23,28]	[4]	[25]	[11]	[28]	[9,19]	[59]		
14	[4]	[ ]	[ ]	[22,48]	[ ]	[93]	[16]	[155]	
15	[9]	[16]	[97]	[6]	[5]	[102]	[18,	[17,55]	[62,66,
16	[4,7]	[6,15]	[12,14]	[10]	[9,24]	[57,58]	[8,14,18,22]	[137,140]	[8,12,
17	[38]	[50]	[23]	[22,59]	[59]	[24,71]	[50,105]	[15,	[
18	[4,9]	[38]	[28]	[6]	[12]	[68,118]	[10,15,16,21]	[	[5]
19	[5]	[9]	[39]	[4,7]	[44]	[13]	[	[11,	[
20	[4]	[92]	[12,26,30]	[61,95]	[5,14,	[46,63]	[8,	[	[10,
21	[9]	[ ]	[9]	[6,8]	[128]	[	[	[	[7]
22	[4]	[61]	[10,14]	[14,17]	[8]	[ ]	[16,30]	[	[9,
23	[13]	[55]	[11,17]	[74,139]	[48,54]	[8,	[	[6,	[
24	[4,9]	[9,11,26]	[12,14]	[6,7]	[11,14,16]	[7,10,17]	[8,	[	[8,10,
25	[46]	[47,83]	[13]	[32]	[5]				

Table 7. Some Known Good Replies after Two Moves. Closed brackets indicate that all good replies are known; [ ] indicates a $\mathcal{P}$-position.

Finally, here is a selection of good replies found by G.L. Sicherman.

Moves	Replies	Moves	Replies	Moves	Replies
{14,18,26,38}	[501]	{6,22,38}	[4,469]	{10,24,26}	[293]
{8,26,28,38}	[205]	{12,16,20,22,26,30}	[183]	{8,34, 36,54}	[173]
{8,22,26}	[10,169]	{10,26,28,32 }	[129]	{14,18,26}	[119,139]
{8,20,34,46}	[26,119]	{12,22,26,30,32}	[119]	{8,26, 30,36}	[12,93,113]

Table 8. Some Good Replies.

3.5 Normal Play and Nim-values.

If the rules are changed so that naming 1 becomes an illegal move then Sylver Coinage becomes a normal-play game and the Sprague-Grundy theory applies. However, with infinitely many options to consider, there is the possibility of infinite nim-values.

For example, the game $\{2, 2n+3\}, n \geq 0$ has as its only options

$$\{2, 2k+3, 2n+3\} \equiv \{2, 2k+3\}, \quad 0 \leq k < n.$$

Now $\mathcal{G}(2,3) = 0$ and inductively

$$\mathcal{G}(2, 2n+3) = \text{mex}\{0, 1, 2, ..., n-1\} = n, \ n \geq 0.$$

The only replies to 2 are odd numbers and so

$$\mathcal{G}(2) = \text{mex}\{0, 1, 2, ...\} = \omega.$$

In contrast to this we find $\mathcal{G}(3) = 1$. To show this we make use of two observations. First, that in a game G if every option of G has an option whose nim-value is k then the options of G cannot have nim-value k and in turn k is a possible nim-value for G. Second, any game in which the minimum named number is 3 has $\{2,3\}$ as an option and since $\mathcal{G}(2,3) = 0$ the nim-value of the game is not 0.

We now proceed inductively. The game $\{3,4,5\}$ has $\{2,3\}$ as its only option and so $\mathcal{G}(3,4,5) = \text{mex}\{0\} = 1$. For $k \geq 1$ the game $\{3, 3k+1, 3k+2\}$ has options $\{2,3\}$ and $\{3, 3j+1, 3k+2\}$, $\{3, 3k+1, 3j+2\}$, $1 \leq j \leq k$. In the latter cases each has $\{3, 3j+1, 3j+2\}$ as an option but, by induction, $\mathcal{G}(3, 3j+1, 3j+2) = 1$. Therefore by our earlier observations it follows that $\mathcal{G}(3, 3k+1, 3k+2) = 1, k \geq 1$. Now every option of $\{3\}$ has an option with nim-value 1, so we in turn see that $\mathcal{G}(3) = 1$.

Find the nim-values of $\{4, 6, 4n-1, 4n+1\}$ (easy) or $\{3, k\}$ (harder) or $\{4\}$ (hardest).

3.6 Comments and Questions.

It is possible to play this game in other contexts. Consider 2-dimensional vectors whose entries are nonnegative integers, and for multiples allow nonnegative integer scalars. When (1,0) and (0,1) have both been named then both normal and misère versions are finished. But, the game does not necessarily end: (1,1), (1,2), (1,3), ... is an infinite sequence of legal moves. Nevertheless, in normal play (1,1) is not a losing move since any move (a, b) can be countered by (b, a).

What is the status of (1,1) in the misère game?

More generally, what about the status of normal and misère play in a vector space with nonegative integers as scalars and appropriate entries in the vectors? The Gaussian integers? (That is, the complex numbers of the form $a + bi$ with a and b integers.) In an integral domain?

4 Dots-and-Boxes

Dots-and-Boxes starts with a rectangular array of dots and two players take turns to join two horizontally or vertically adjacent dots. If a player completes the fourth side of a unit square (**box**) he initials that box and must then draw another line. A player who can complete a box is not obliged to do so if he has something else he prefers to do. When all the boxes have been completed the game ends and whoever has initialled more boxes is declared the winner. In this form the game is not one to which the Sprague-Grundy theory immediately applies.

Play would become simpler were a player obliged to take a box when one became available. J.C. Holladay(1966) has shown that this version of the game can be easily analyzed provided that when the stage has been reached that boxes have to be given away then the rectangular array is composed of long chains and cycles. In this case

$$\begin{aligned}\text{the number of moves} &= \text{total number of edges to be drawn}\\ &\quad - \text{the total number of squares}\\ &\quad + \text{number of plays that complete two squares} + 1.\end{aligned}$$

The last concept, 'a play that completes two squares' is important in the theory that follows. It is called a **double-cross**.

Dots and Boxes is, like other good games, remarkable in that it can be played on several different levels of sophistication. As a guide to what is to come, here is a brief summary.

First Level: *Don't give away a box unless forced to; then give as few as possible.*

This seems to be the only level that many players reach.

Second Level: *Use the Double-Dealing endgame technique.*

This technique gets the winner a lot of boxes at the finish and makes it seem likely that a nim-like game—Nimstring—will be useful.

Third Level: *Use the Parity Rule for long chains.*

To get the right parity is an exercise in Sprague-Grundy theory, so we need tables of nim-values for Nimstring. The unwieldiness of these tables forces us to use equivalence theorems whenever we can, and to look for interesting classes of analyzable graphs. Twopins theory can be used to reduce many games to manageable positions.

Professional Level: *Beyond Nimstring theory.*

4.1 First level

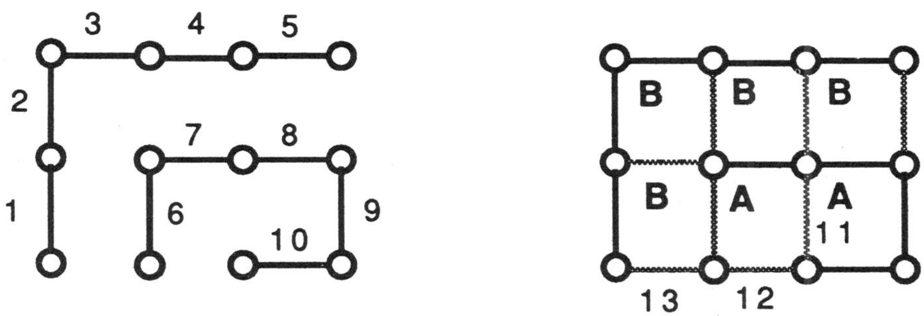

Figure 4: Betty Wins the 2 by 3 Game.

Figure 4 shows a typical first game, in which Albert started. Nothing was given away in the fairly typical opening until move 11. Albert took 2 boxes. His last bonus move enabled Betty to take the last 4. Generally, this is how most people play: boxes are taken as soon as they are available and the bonus move is used to give away the next smallest string of boxes. Clearly, this is not always a winning strategy.

4.2 Second level

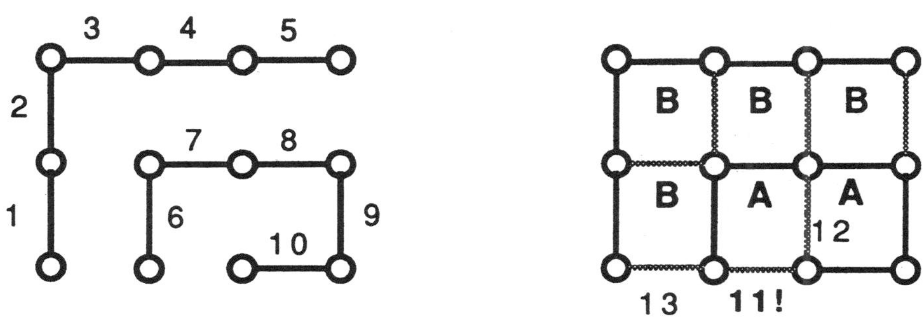

Figure 5: Betty Still Wins the 2 by 3 Game!

Most children (and adults) play only at the first level, but Betty is a Second Level player. She started the return match with the same opening that Albert had used. He was happy to copy Betty's replies from that game and was delighted to see her follow it even as far as that tenth move, which had proved his undoing (Figure 5). Instead of taking the two boxes, Betty astounded him with move 11. He was doubled-crossed! He now has the choice of losing 4 to 2 or 6 to 0.

You can beat almost everyone in this double-dealing way i.e.

when your opponent opens a long chain for you, close it off with a double-dealing move which gives your opponent the last 2 boxes but forces him to open the next chain for you.

In this way you keep control right to the end of the game. Each double-dealing move is followed, usually immediately, by a move in which two boxes are completed in one turn. These moves are called **doublecrossed** moves (see Figure 6), because, to quote *Winning Ways*, 'whoever makes them usually has been'.

Figure 6: A Double-Dealing and a Double-Crossed move.

This strategy suggests the following policy:

Make sure that there are long chains about
and try to force your opponent to be the first to surrender one.

We'll say that whoever can force her opponent to surrender a long chain has **control** of the game. The player who has control usually wins decisively when there are several long chains. Two second level opponents would soon find themselves reaching for the

4.3 Third Level

Getting and keeping control is now the paramount concern. This in turn depends on the number of dots, the number of double-crossed moves and whether you are playing first or second. In particular you should observe

The Parity Rule:

Initial dots + doublecrossed moves should be:
odd, if you play first and **even**, if you play second.

In simple games the number of doublecrosses will be one less than the number of long chains and this rule becomes:

The Long Chain Rule:

Try to make the number of initial dots + eventual long chains
even, if you move first and **odd**, if you move second.

(the **opposite** parity from your playing number).

The reason for these rules is that whatever shape board you have on your paper, you'll find that

Number of initial dots + Number of doublecrosses = Total number of turns in the game.

Can you prove this?

The problem with this strategy is the proper definition of a long chain. It should be a chain in which a doublecrossed move can occur. Therefore a long chain contains 3 or more squares and a short chain contains 2 squares or 1 .

In using this strategy, note that a closed cycle of 4 or more cells counts as two long chains because you must offer your opponent **two** doublecrossed moves, i.e. four instead of two boxes. Note also that each doublecross, no matter who makes it, changes the number of long chains you want. So as a rule of thumb, make the long chains as long as possible so that you gain overall from a doublecross move and also try to avoid cycles unless the chains are very long. It is clear that if the long chains are mainly of length 3 and 4 it is possible to retain control throughout the game yet still lose.

4.4 Professional Level

Double-dealing does not help when the majority of the long chains are short (3 or 4 in length). The author is still an amateur and does not dare to tangle with the 'burly one'.

4.5 Strings-and-Coins and Nimstring

To aid the Third Level players in their fight for control, we examine a dual form of Dots-and-Boxes, called **Strings-and-Coins.** The ends of each piece of string are connected to two different coins or to a coin and the 'ground'. Each player in turn cuts a string. If a cut completely detaches a coin, the player takes the coin and must then cut another string (if there's one still uncut). The game ends when all coins are detached and the player who pockets the greater number of coins is the winner.

It's easy to see that for any Dots-and-Boxes position there is a corresponding Strings-and-Coins position (coins are placed in the centers of the squares and each string crosses exactly one (potential) edge in the original diagram, the ground being the outside of the rectangular array). However, playing Strings-and-Coins on a non-planar graph would not correspond to any Dots-and-Boxes position so Strings-and-Coins is a generalization of Dots-and-Boxes. (We leave the thrill of experimentation and discovery for the reader.) **Nimstring** is a variant of Strings-and-Coins. The moves are the same **but**

WARNING. In Nimstring you **lose** when you detach the last coin, for then the rules require you to make a further move when it is impossible to do so.

We still want to use double-crossing as the basis for our strategies in these games. Now a **short** chain is still one with two or less coins, but a **short** cycle has three or less coins. A cycle of three could not happen in Dots-and-Boxes. Anything that is not short is **long.**

Although the two games look quite different, Nimstring is a special case of Strings-and-Coins. To show this, let G represent an arbitrary Nimstring position, add a long chain (with more coins in it than there are in G), grounded at both ends, and consider the resulting Strings-and-Coins game. Because the chain is so long, both players will try to avoid cutting any string of it. Neither player can force his opponent to move on the chain until all the strings of G have been cut. In other words, the only way to win the Strings-and-Coins game is to play a winning game of Nimstring on the graph G.

Another construction using the **double-crossing strategy** is to add several long chains and cycles to the Nimstring game G to get a Strings-and-Coins game. If these are long enough the winning strategy for the Strings-and-Coins game is then: If your opponent moves in G, reply in G with a move from the winning Nimstring strategy. If he moves in a long chain, take all but two coins of that chain, leaving just the string which joins them. If he moves in a long cycle, take all but four coins of the cycle, leaving them as two pairs each joined by a string.

This strategy gives you all but 2 coins of each long chain and all but 4 of each long cycle, so it will win for you if the total number of nodes in the added chains and cycles exceeds (the number of nodes in G) + 4 × (the number of added long chains) + 8 × (the number of added long cycles).

The conclusion therefore is:

To play Strings-and-Coins and Dots-and-Boxes you must know how to play Nimstring.

4.6 Sprague-Grundy Theory for Nimstring

The discussion of nim-values for Nimstring is clouded by situations such as the following: A coin joined to another coin which in turn is connected to the ground. Suppose that the value for this position, call it K, is the nimber x. Now any game composed of graph $G + K$ is a first player win which can be demonstrated by strategy-stealing. (**Do it**) Therefore, for every nimber y, including x itself, we have $x \overset{*}{+} y \neq 0$ and in particular $x \overset{*}{+} x \neq 0$. This is called a **loony** position, whose value is $\bowtie$ (think of it as $-\infty$). The theory of complementing and complementary moves in Chapter 12 of *Winning Ways* applies to Nimstring (where the complementing moves are those that capture coins) and shows that every position has either an ordinary nim-value or the special value $\bowtie$. However, the entire theory is not needed since the rules for finding values at Nimstring can be deduced after examining the six positions in which a coin may be captured on the next move.

(a) A coin joined to the ground.	TAKE.	**A free coin.**
(b) Two joined coins.	TAKE.	**Two free coins and a doublecross.**
(c) A coin joined to two others.	TAKE.	**Three free coins and a doublecross.**
(d) A coin joined to a coin which is joined (multiply) to *G*.	TAKE.	**A free coin.**
(e) A coin joined to the ground and another coin.	WIN!	
(f) A coin joined to two coins, one of which is joined (multiply) to *G*.	WIN!	

In the first four cases, if there is a winning strategy for the next player which involves cutting some edge X she may as well first take the coins then cut X. The case (e) *you* analyzed earlier. Situation (f) is similar to (e) and also uses strategy-stealing.

The values of Nimstring positions are:

The value of a graph without strings is 0.

The value of a graph with a capturable coin of one of the four types (a)-(d) is equal to that of the subgraph obtained by removing the capturable coin(s) and its string(s).

The value of a graph with a capturable coin of one of the two types (e) and (f) is $\bowtie$.

The value of a graph with no capturable coins is found from the values of the graphs left after cutting single strings by using the Mex Rule.

When calculating with these values, remember that

$$\bowtie \overset{*}{+} 0 = \bowtie \overset{*}{+} 1 = \bowtie \overset{*}{+} 2 = \ldots = \bowtie \overset{*}{+} \bowtie = \bowtie,$$

as well as the ordinary nim-addition rules.

Using these rules, show that the Nimstring version of the 3 by 3 Dots-and-Boxes game is a second player win.

There are many techniques which help to reduce Nimstring graphs down to a more manageable size which we will only mention here. For example, in a long chain with no edges to the ground an edge can be replaced by two edges going to the ground provided that this leaves two shorter but technically long chains. This is called **snapping.** For more detail the reader should consult *Winning Ways*, 530-535.

Many times the subgraphs have components that are without circuits or capturable nodes in which all the **joints** (nodes of valence three or more) lie on a single long path (the stem) and each joint belongs to just 3 edges. Such a component is called a **vine.** The chain joining an end to its nearest joint is called a **tendril**, so a single-jointed vine has 3 tendrils. Vines with more joints have 2 tendrils at their endmost joints and just 1 at intermediate ones. If the distance between two neighboring joints is long, the vine decomposes into two smaller ones because long chains snap, so we can suppose such distances short, if we like.

A **Twopins-vine** is one whose every distance between non-neighboring stops (which may be either ends or joints) is long. We call it a Twopins-vine because of the remarkable fact that the value of any such vine is equal to that of a corresponding configuration in the game of

4.7 Twopins

This is a game like Kayles, played with a row of skittles, except that the skittles may now occur in pairs, one behind another. As in Kayles, you can knock down the skittles in two adjacent columns, or in one column provided it has two skittles in it. A complete analysis of the game of Twopins is beyond our grasp, but special cases have been shown to be equivalent to Dawson's Kayles and to Kayles itself. Enough information to enable you to play any reasonable sized game of Dots-and-Boxes is given in *Winning Ways*, 466–467.

Each joint in a Twopins-vine with a short tendril is equivalent to a column of two pins; each other joint to a column of one; and two neighboring joints a long distance apart correspond to an empty Twopins column. A bowling shot which removes a single column at Twopins corresponds to a tendril move at Nimstring; one which removes a pair of columns corresponds to a move on the stem of the vine.

4.8 Betty beats Albert a third time!

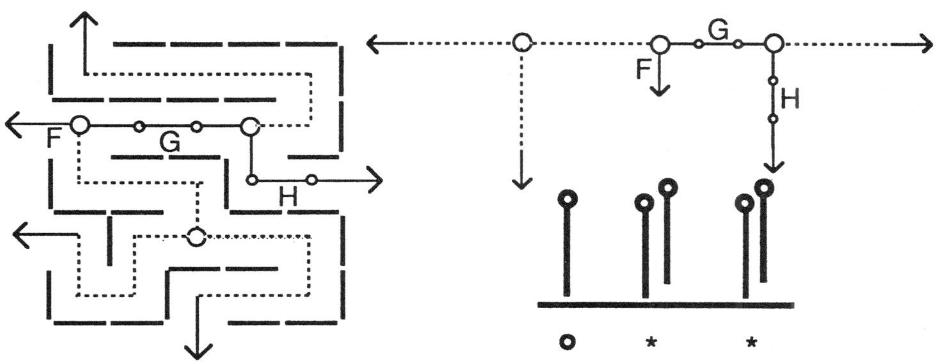

Figure 7: Betty beats Albert at the Third Level.

Figure 7 shows a 5 by 5 game of Dots-and-Boxes in which Albert played first and has just made his 16th move. He thinks that Betty will make the Free move F to which he will proudly reply with the Hard-hearted handout H! But Betty was already aware that the position is equivalent to the Nimstring position in Figure 6, which in turn is equivalent to the Twopins position whose value is $**$ and $\circ$, i.e., $2 \overset{*}{+} 0 = 2$. She boldly knocks down **both** columns of twopins by making the Good move G!

Notice that Betty doesn't draw the vertical edge to the immediate left or right of G, since this would give Albert a chance to reply with the other of these two edges—a hard-hearted handout.

The number of dots in this example is 36, the eventual number of long chains is three and the number of double-crosses will be 2, illustrating the Parity Rule and the Long Chain Rule given at Level Three. G is the 32nd move; they will each have three more turns, Albert surrendering a chain each time, and having the doubtful privilege of making the two double-crossing moves. A total of 38 turns, illustrating the "Dots + Doublecrosses = Turns" theorem.

4.9 Comments and Questions

There are three possible 3-dimensional versions of Dots-and-Boxes. One, draw edges and claim unit-squares; two, draw edges but claim unit-cubes; and three, draw unit-squares and claim unit-cubes. Version three can be dualized and Nimstring theory helps.

What about the other two?

5 Hints and Answers

The **Sliding Game** with $\omega+1$ strips can be played as this 'almost'disjunctive compound except it needs **three** new strips — the ω and $\omega+1$ strips plus a ω^2 strip to account for the 'infinities' of these two boards.

The nim-value of (2, 4, 7, 8, 10, 13, 14, 16) in **Welter's Game** is 16 and the only good move is from 16 to 0.

You shouldn't want to play first from the position $\{\omega, \omega+1\}$ in Welter's Game, because the first player must go to $\{x, \omega\}$ or to $\{x, \omega+1\}$ and the good reply by the second player is to $\{x-1, x\}$ or to $\{x, x+1\}$. Which is it?

From $\{\omega, \omega+1, \omega+2\}$ the first player can win by playing from $\omega+2$ to 0, but she should note the parity change when it comes to her next move. Why doesn't the move ω to 0 win?

As you saw soon after, the good reply to 4 in **Sylver Coinage** is 6.

The unique good reply to $\{5, 6\}$ in Sylver Coinage is 19.

The nim-value of $\{4, 6, 4n-1, 4n+1\}$ in Sylver Coinage is 1 for all values of n.

The nim-value $\mathcal{G}(3, k)$ for $k =$ 4, 5, 6, 7, 8, 9, 10, 11, 12, 13, 14, 15, 16 ...
is 2, 3, 1, 4, 6, 1, 7, 8, 1, 9, 11, 1, 12 ...,

respectively. The general pattern is unknown.

The value $\mathcal{G}(4)$ is unknown, perhaps it is $\omega+1$? Or 6?

We don't know the status of (1, 1) in two-dimensional misère Sylver Coinage.

You can prove the last theorem that we mentioned about **Dots-and-Boxes** by using Euler's formula for planar maps. You can also use Euler's formula to calculate the number of moves in **Brussels Sprouts** until there is just one arm in each region. (See M. Gardner, Mathematical Games: of Sprouts and Brussels Sprouts, games with a topological flavor, *Scientific Amer.*, **235**#1 (Jul. 1967) 112-115.) It turns out that Brussels Sprouts is just a clever disguise for **She-Loves-Me-She-Loves-Me-Not**!

Proceedings of Symposia in Applied Mathematics
Volume **43**, 1991

Unsolved Problems in Combinatorial Games

Richard K. Guy

1. Subtraction games are known to be periodic. Investigate the relationship between the subtraction set and the length and structure of the period. The same question can be asked about *partizan* subtraction games, in which each player is assigned an individual subtraction set. See Fraenkel & Kotzig (1987).

[**Subtraction games** are played with heaps of beans. A move is to take a number of beans from a heap, provided that number is a member of the **subtraction set**. See Subtraction Games in WW, 83–86, 487–498 and in Section 4. of the Impartial Games chapter of this book.]

2. Are all finite octal games ultimately periodic? Resolve any number of outstanding particular cases, e.g., **·6** (Officers), **·06**, **·14**, **·36**, **·64**, **·74**, **·76**, **·004**, **·005**, **·006**, **·007** (One-dimensional tic–tac–toe, Treblecross), **·016**, **·104**, **·106**, **·114**, **·135**, **·136**, **·142**, **·143**, **·146**, **·162**, **·163**, **·172**, **·324**, **·336**, **·342**, **·362**, **·371**, **·374**, **·404**, **·414**, **·416**, **·444**, **·454**, **·564**, **·604**, **·606**, **·644**, **·744**, **·764**, **·774**, **·776** and Grundy's Game (split a heap into two unequal heaps). Find a game with a period longer than 149459. Explain the structure of the periods of games known to be periodic.

[If the binary expansion of the kth code digit in the game with code $\mathbf{d}_0 \cdot \mathbf{d}_1\mathbf{d}_2\mathbf{d}_3 \ldots$ is

$$\mathbf{d}_k = 2^{a_k} + 2^{b_k} + 2^{c_k} + \ldots,$$

where $0 \leq a_k < b_k < c_k < \ldots$, then it is legal to remove k beans from a heap, provided that the rest of the heap is left in exactly a_k or b_k or c_k or $\ldots$ non-empty heaps. See WW, 81–115 and Gangolli & Plambeck (1989) and Section 4. of the Impartial Games chapter of this book.]

The paper is in final form and no version of it will be submitted for publication elsewhere.
1980 *Mathematics Subject Classification* (1985 *Revision*). Primary 90D05.

3. Examine some hexadecimal games. Obtain conditions for arithmetic periodicity.

[**Hexadecimal games** are those with code digits $\mathbf{d}_k$ in the interval from **0** to **15**. See WW, 115–116.]

4. Extend the analysis of **Domineering** to larger boards. For a modest beginning, find the values of 4×4 and 4×5 boards.

[Left & Right take turns to place dominoes on a checker-board. Left orients her dominoes North-South and Right orients his East-West. Each domino exactly covers two squares of the board and no two dominoes overlap. A player unable to play loses. See Berlekamp (1988); WW, 495–498; and Section 9. of the first chapter of this book.]

5. Analyze positions in the game of **Go**.

[Compare Berlekamp (1988), and his continuing discoveries, discussed in the present volume.]

6. Is **Go-Moku** (Five-in-a-row, Go-Bang, Pegotty) a win for the first player?

7. Complete the analysis of impartial **Eatcakes** (WW, 269, 271, 276–277).

[Played with a number of rectangles, $m_i \times n_i$; a move is to remove a strip $1 \times n_i$ or $m_i \times 1$ from each rectangle, either splitting it into two rectangles, or reducing the length or breadth by one. Winner removes the last strip.]

8. Complete the analysis of **Hotcakes** (WW, 279–282).

[Also played with integer-sided rectangles. Left cuts as many rectangles vertically along an integer line as she wishes, and then rotates one of each pair of resulting rectangles through a right angle. Right cuts as many rectangles as he wishes, horizontally into pairs of integer-sided rectangles and rotates one rectangle from each pair through a right angle.]

9. Develop a misère theory for unions of partizan games.

[In a union of two or more games, you move in as many component games as you wish. In misère play, the last player *loses*.]

10. Extend the analysis of **Squares Off** (WW, 299).

[Played with heaps of beans. A move is to take a perfect square (> 1) number of beans from any number of heaps. Heaps of 0, 1, 2 or 3 cannot be further reduced. A move leaving a heap of 0 is an overriding win for the player making it. A move leaving 1 is an overriding win for Right, and one leaving 2 is an overriding win for Left. A move leaving 3 doesn't end the game unless all other heaps are of size 3, in which case the last player wins.]

11. Extend the analysis of **Top Entails** (WW, 376–377).

[Played with stacks of coins. Either split a stack into two smaller ones, or remove the top coin from a stack. In the latter case your opponent's move must use the same stack. Last player wins. Don't leave a stack of 1 on the board, since your opponent must take it and win, since it's now your turn to move in an empty stack!]

12. Extend the analysis of **All Square** (WW, 385).

[Played with heaps of beans. A move splits a heap into two smaller ones. If both heap sizes are perfect squares, the player must move again: if he can't he loses!]

13. Extend the misère analysis of various octal games, e.g., **Officers**, **Dawson's Chess**, ..., and of **Grundy's Game**.

[See Allemang (1984), WW, 411–421. Recently, William L. Sibert completed the analysis of misère Kayles; see the post-script to Sibert & Conway (1990).]

14. Moebius, when played on 18 coins has a remarkable pattern. Is there any trace of pattern for larger numbers of coins? Can any estimate be made for the rate of growth of the nim-values?

[See Coin-turning games in WW, 432–435; Vera Pless's lecture; and Section 8. of the Impartial Games chapter of this book. Played with a row of coins. A move turns 1, 2, 3, 4 or 5 coins, of which the rightmost must go from heads to tails. Winner is the player who makes all coins tails.]

15. Mogul has an even more striking pattern when played on 24 coins, which has some echoes when played on 40, 56 or 64 coins. Thereafter, is there complete chaos?

[See references for problem 14. A move turns 1, 2, ..., 7 coins.]

16. Find an analysis of **Antonim** with four or more coins (WW, 459–462).

[Played with coins on a strip of squares. A move moves a coin from one square to a smaller-numbered square. Only one coin to a square, except that square zero can have any number of coins.]

17. Extend the analysis of **Kotzig's Nim** (WW, 481–483). Is the game eventually periodic in terms of the length of the circle for every finite move set? Analyze the misère version of Kotzig's Nim.

[Players alternately place coins on a circular strip, at most one coin on a square. Each coin must be placed m squares clockwise from the previously placed coin, provided m is in the given **move set**, and provided the square is not already occupied. The complete analysis is known only for a few small move sets.]

18. Obtain asymptotic estimates for the proportions of $\mathcal{N}$-, $\mathcal{O}$- and $\mathcal{P}$-positions in Epstein's **Put-or-Take-a-Square** game (WW, 484–486).

[Played with one heap of beans. At each turn there are just two options, to add or take away the largest perfect square number of beans that there is in the heap.]

19. Simon Norton's game of **Tribulations** is similar to Epstein's game, but squares are replaced by triangular numbers. Norton conjectures that there are no $\mathcal{O}$-positions, and that the $\mathcal{N}$-positions outnumber the $\mathcal{P}$-positions in golden ratio. True up to 5000 beans.

20. Complete the analysis of **D.U.D.E.N.E.Y.**

[Played with a single heap of beans. Either player may take any number of beans from 1 to Y, except that the immediately previous move mustn't be repeated. When you can't move you lose. Analysis easy for Y even, and known (WW, 487–489) for 53/64 of the odd values of Y.]

21. **Schuhstrings** is the same as D.U.D.E.N.E.Y., except that a deduction of zero is also allowed, but cannot be immediately repeated (WW, 489–490).

22. Analyze Nim in which you are not allowed to repeat a move. There are at least five possible forms (assume that b beans have been taken from heap H):

medium local: b beans may not be taken from heap H until some other move is made in heap H.

short local: b beans may not be taken from heap H on the next move.

long local: b beans may never again be taken from heap H.

short global: b beans may not be taken from any heap on the next move.

long global: b beans may never again be taken from any heap.

23. Analyze Nim played with a row of heaps. A move may only be made in the leftmost or in the rightmost heap. When a heap becomes empty, then of course its neighbor becomes the extremal heap.

24. Continue the analysis of **The Princess and the Roses** (WW, 490–494).

[Played with heaps of beans. Take one bean, or two beans, one from each of two different heaps.]

25. Analyze the Conway-Paterson game of **Sprouts** with 7 or more spots, or the misère form with 5 or more spots (WW, 564–568).

[A move joins two spots, or a spot to itself by a curve which doesn't meet any other spot or previously drawn curve. When a curve is drawn, a new spot must be placed on it. The valence of any spot must not exceed three. Since this question was asked, Daniel Sleator of A.T.&T. Bell Laboratories has pushed the normal analysis to about 10 spots and the misère analysis to about 8.]

26. Extend the analysis of **Sylver Coinage** (WW, 575–597).

[Players alternately name different positive integers, but may not name a number which is the sum of previously named ones, with repetitions allowed. Whoever names 1 loses. See Section 3. of the last chapter in this book.]

27. Extend the analysis of **Chomp** (WW, 598–599).

[Players alternately name divisors of N, which may not be multiples of previously named numbers. Whoever names 1 loses. David Gale offers a prize of US$100.00 for the first complete analysis of 3D-Chomp, i.e., where N has three distinct prime divisors, raised to arbitrarily high powers.]

28. Extend Uléhla's or Berlekamp's analysis of von Neumann's game from diforests to directed acyclic graphs (WW, 570–572; Uléhla [1980]). Since Chomp and the superset game [Gale & Neyman (1982)] can be described by directed acyclic graphs but not by diforests, the proposed extension could presumably throw some light on these two unsolved games.

29. Prove that Black doesn't have a forced win in Chess.

30. Simon Norton's King and Rook versus King problem. Played on a quarter-infinite board, with initial position WKa1, WRb2 and BKc3. Can White win? If so, in how few moves? It may be better to ask, "what is the smallest board (if any) that White can win on if Black is given a win if he walks off the North or East edges of the board?" Is the answer 9×11?

31. David Gale's version of Lion and Man. L and M are confined to the non-negative quadrant of the plane. They move alternately a distance of at most one unit. For which initial positions can L catch M?

Conjecture: L wins if he is northeast of M. This condition is clearly necessary, but the answer is not known even if M is at the origin and L is on the diagonal.

Variation. Replace quadrant by wedge-shaped region.

32. Gale's Vingt-et-un. Cards numbered 1 through 10 are laid on the table. L chooses a card. Then R chooses cards until his total of chosen cards exceeds the card chosen by L. Then L chooses until her cumulative total exceeds that of R, etc. The first player to get 21 wins. Who is it?

33. Subset Take-away. Given a finite set, players alternately choose proper subsets subject to the rule that once a subset has been chosen, none of *its* subsets may be chosen subsequently by either player. Last player wins.

[David Gale conjectures that it's a second player win—this is true for sets of less than six elements.]

34. R.B. Eggleton & A.S. Fraenkel ask for a theory of **Cannibal Games** or an analysis of special families of positions. They are played on an arbitrary finite digraph. Place any numbers of "cannibals" on any vertices. A move is to select a cannibal and move it along a directed edge to a neighboring vertex. If this is occupied, the incoming cannibal eats the whole population (**Greedy Cannibals**) or just one cannibal (**Polite Cannibals**). A player unable to move loses. Draws are possible. A partizan version can be played with cannibals of two colors, each eating only the opposite color.

35. Welter's Game on an arbitrary digraph. Place a number of monochromatic tokens on distinct vertices of a directed acyclic graph. A token may be moved to any *unoccupied* immediate follower. Last player wins. Formulate a winning strategy.

36. Restricted Positrons & Electrons. Place Positrons (Pink tokens) and Electrons (Ebony tokens) on some distinct vertices of a Welter strip. Any particle can be moved by either player leftward to any square u provided that u is either unoccupied or occupied by a particle of the opposite type. In the latter case, of course, both particles become annihilated (i.e., they are removed from the strip), as physicists tell us positrons and electrons do. Play ends when the excess particles of one type over the other are jammed in the lowest positions of the strip. Last player wins. Formulate a winning strategy. Note that if the particles are of one type only, this is Welter's Game. As a strategy is known for Misère Welter [WW, 480–481] it may not be unreasonable to ask for a misère analysis as well.

37. General Positrons & Electrons. As 36. but played on an arbitrary digraph. Last player wins.

Proceedings of Symposia in Applied Mathematics
Volume **43**, 1991

Selected Bibliography on Combinatorial Games and Some Related Material

Aviezri S. Fraenkel

Department of Applied Mathematics & Computer Science
The Weizmann Institute of Science
Rehovot 76100, Israel

1. What are Combinatorial Games? What are they Good For?

Roughly speaking, the family of *combinatorial games* consists of 2-player games with perfect information (no hidden information as in some card games), no chance moves (no dice) and outcome restricted to (lose,win), (tie,tie) and (draw,draw) for the 2 players who move alternately. Tie is an end position such as in tic-tac-toe, where no player wins, whereas draw is a dynamic tie, i.e., any position from which a player has a nonlosing move, but cannot force a win. Note that chess is a combinatorial game. We use the shorter terms *game* and *games* below to designate combinatorial games.

Amusing oneself with games may sound like a frivolous occupation. But the fact is that the bulk of interesting and natural mathematical problems which are hardest in complexity classes beyond *NP*, such as *P*space, *E*xtime and *E*xspace, are 2-player games. Two of the reasons for this high complexity are outlined below. Before

The paper is in final form and no version of it will be submitted for publication elsewhere.
1980 *Mathematics Subject Classification* (1985 *Revision*). Primary 90D05.

that we note that in addition to a natural appeal of the subject, there are applications or connections to various areas, including complexity, logic, graph and matroid theory, networks, coding theory and surreal numbers.

2. Why are Combinatorial Games Hard?

Decision problems such as Traveling Salesperson (Is there a round tour through specified cities of cost $\leq C$?) or graph hamiltonicity are *existential*: "Is there...?", involving a single existential quantifier. In mathematical terms an existential problem boils down to finding a path, in a large "decision-tree" of all possibilities, which satisfies specified properties. The above 2 problems, as well as thousands of other interesting and important combinatorial-type problems are NP-*complete*. This means that they are *conditionally intractable*, i.e., the best way to solve them seems to require traversal of most if not all of the decision tree, whose size is exponential in the input size of the problem. No better method is known to date at any rate.

The decision problem whether White can win if White moves first in a chess game, on the other hand, has the form: "$\exists\forall\exists\forall\cdots$ moves: White wins?", i.e., is there a move of White such that for every move of Black there is a move of White$\cdots$ such that White can win? Here we have a large number of alternating quantifiers rather than a single existential one. We are looking for an entire subtree rather than just a path in the decision tree. The problem for generalized chess on an $n \times n$ board is in fact Extime-complete, which is a *provable intractability*. Most games are at least Pspace-hard.

Put in simple language, in analyzing an instance of Traveling Salesperson, the problem itself is passive: it does not resist your attempt to attack it, yet it is difficult. In a game, in contrast, there is your opponent, who, at every step, attempts to foil your effort to win. It's similar to the difference between an autopsy and surgery. Einstein, contemplating the nature of physics said, "Der Allmächtige ist nicht böshaft; Er ist raffiniert" (The Almighty is not mean; He is sophisticated). NP-complete existential problems are perhaps sophisticated. But your opponent in a game can be very mean!

Another reason for the high complexity of games is connected with the fundamental notion of *sum* (disjunctive compound) of games. A sum is a finite collection of disjoint games. Each of the 2 players at his turn selects one of the games and makes a move in it. If the outcome is not a draw, the sum-game ends when there is no move left in any of the component games. If the outcome is not a tie either, then in *normal* play, the player first unable to move loses and his opponent wins. The outcome is reversed in *misère* play.

The *game-graph* of a game is a directed graph whose vertices are the positions of the game, and (u, v) is an edge if and only if there is a move from position u to position v. It turns out that the game-graph of a sum has size exponential in the combined size of the input game-graphs! Since sums occur naturally and frequently, and since analyzing the sum entails reasoning about its game-graph, we are faced with a problem which is a priori exponential, quite unlike most present-day interesting existential problems.

3. Breaking the Rules

As the experts know, some of the most exciting games are obtained by breaking some of the rules for combinatorial games, e.g., permitting a player to pass a bounded or unbounded number of times, relaxing the requirement that players play alternately, or permitting a number of players different from 2. But by far the most fruitful tampering with the rules seems to be to permit sums of games which are not quite fixed (which explains why misère play of sums of games is much harder than normal play) or not quite disjoint (Welter) or the game does not quite decompose into a sum (Wythoff).

On the other hand, permitting a payoff function different from (0,1) = (lose,win) or (1/2,1/2) = (draw,draw) usually leads to games which are not considered to be combinatorial games, or to borderline cases.

4. Why is the Bibliography Vast?

In the realm of existential problems, such as sorting or Traveling Salesperson, most present-day interesting decision problems can be classified into tractable, conditionally intractable, and provably intractable ones. There are exceptions, to be sure, e.g. graph isomorphism and primality testing, whose complexity is still unknown. But these are few. It appears that, in contrast, there is a very large set of games whose complexities are hard to determine. The set of these games is termed *Wonderland*, because we are wondering about the complexity classification of its members. Only few games have been classified into the complexity classes they belong to. Today, most games belong to Wonderland, and despite recent impressive progress, the tools for reducing Wonderland are still few and inadequate.

To give an example, many interesting games have a very succinct input size, so a polynomial strategy is often more difficult to come by (Octal games; Grundy's game). Succinctness and non-disjointness of games in a sum may be present simultaneously (poset games). In general, "breaking the rules" and the alternating quantifiers add to the volume of Wonderland. I suspect that the large size of Wonderland, a fact of independent interest, is the main contributing factor to the bulk of the bibliography on games.

5. Why isn't it Larger?

The bibliography below is a partial list of books and articles on combinatorial games and related material. It is partial not only because I constantly learn of additional relevant material I did not know about previously, but also because of certain self-imposed restrictions. The most important of these is that only papers with some original and nontrivial mathematical content are considered. This excludes most, but not all, of the work on heuristic or artificial intelligence approaches to games, especially the large literature concerning computer chess. It also excludes most historical reviews of games.

On the other hand, papers on games which break some of the rules of combinatorial games are included liberally, as long as they are interesting and retain a combinatorial flavor. These are vague and hard to define criteria, yet combinatorialists usually recognize a combinatorial game when they see it. Besides, it is interesting to break also this rule sometimes! Adding borderline cases is acknowledged in the "related material" postfixed to the title of this bibliography. We have also included papers on various applications of games, especially when the connection to games is substantial.

On the administrative side, Technical Reports have been excluded (there are 5 exceptions) as well as submitted papers (two exceptions at present), but papers in press and a small number of unpublished theses are included.

The bibliography is very dynamic in nature. Previous versions have been circulated to colleagues for many years, since 1983 I think, at intermittent time intervals. Prior to every mailing updates were prepared, and usually also afterwards, as a result of the comments received from several correspondents. The listing can never be "complete". Thus also the present form of the bibliography, published in these Proceedings at the solicitation of Richard Guy, is by no means complete. In fact, I wish to take this opportunity to ask the readers to send me corrections and inform me of gross omissions, remembering, however, that it is a *selected* bibliography. I prefer to get reprints or preprints rather than only titles, whenever possible. On the other hand, there may soon be a *final incomplete* edition, as the funds for producing and distributing this list are dwindling fast.

Most of the bibliographic entries refer to items written in English, though there is a sprinkling of Danish, Dutch, French, German, Japanese and Slovakian, as well as some English translations from Russian. The predominance of English may be due to certain prejudices, but it also reflects the fact that nowadays the *lingua franca* of science is English. In any case, I'm soliciting also papers in languages other than English, especially if accompanied by an abstract in English.

Three correspondents, including myself, have expressed the opinion that the value of the bibliography would be enhanced if it would be transformed into an annotated bibliography. This may be done sometime in the future.

Abbreviations of book series and journal names follow the *Math Reviews* conventions. Another convention is that de Bruijn appears under D, not B; von Neumann under V, not N, McIntyre under M not I, etc.

6. Acknowledgments

Many people suggested additions to the bibliography, or contributed to it in other ways. Among them are: Akeo Adachi, Ingo Althöfer, Thomas Andreae, Adriano Barlotti, József Beck, Claude Berge, Gerald E. Bergum, H.S. MacDonald Coxeter, Thomas S. Ferguson, James A. Flanigan, Fred Galvin, Martin Gardner, Alan J. Goldman, Solomon W. Golomb, Richard K. Guy, Shigeki Iwata, David S. Johnson, Victor Klee, Donald E. Knuth, Anton Kotzig, Jeff C. Lagarias, Michel Las Vergnas, Hendrik W. Lenstra, Hermann Loimer, F. Lockwood Morris, Richard J. Nowakowski, Judea Pearl, J. Michael Robson, David Singmaster, Cedric A.B. Smith, Rastislaw Telgársky, Yōhei Yamasaki and many others. Thanks to all and keep up the game! Special thanks are due to Ms. Sara Fliegelmann and Mrs. Carol Weintraub, who have been maintaining and updating the bibliography-file, expertly and skilfully, over several different TeX generations.

7. The Bibliographic Listing

1. A. Adachi, S. Iwata and T. Kasai [1981], Low level complexity for combinatorial games, Proc. 13th Annual ACM Symp. Theory of Computing (Milwaukee, WI, 1981), Assoc. Comput. Mach., New York, NY, pp. 228–237.

2. A. Adachi, S. Iwata and T. Kasai [1984], Some combinatorial game problems require $\Omega(n^k)$ time, *J. Assoc. Comput. Mach.* **31**, 361–376.

3. H. Adachi, H. Kamekawa and S. Iwata [1987], Shogi on $n \times n$ board is complete in exponential time, *Trans. IEICE* **J70-D**, 1843–1852 (in Japanese).

4. W. Ahrens [1910], *Mathematische Unterhaltungen und Spiele*, Zweite vermehrte und verbesserte Auflage, Vol. I, Teubner, Leipzig.

5. M. Aigner and M. Fromme [1984], A game of cops and robbers, *Discrete Appl. Math.* **8**, 1–12.

6. M. Ajtai, L. Csirmaz and Zs. Nagy [1979], On a generalization of the game Go-Moku I, *Studia Scientiarum Math. Hungar.* **14**, 209–226.

7. D.T. Allemang [1984], Machine computation with finite games, M.Sc. thesis, Cambridge University.

8. I. Althöfer [1988], Nim games with arbitrary periodic moving orders, *Internat. J. Game Theory* **17**, 165–175.

9. I. Althöfer [1988], On the complexity of searching game trees and other recursion trees. *J. Algorithms* **9**, 538–567.

10. I. Althöfer [1989], Generalized minimax algorithms are no better error correctors than minimax is itself, in: *Advances in Computer Chess* **5** (D.F. Beal, ed.), Elsevier, Amsterdam, pp. 265–282.

11. T. Andreae [1984], Note on a pursuit game played on graphs, *Discrete Appl. Math.* **9**, 111-115.

12. T. Andreae [1986], On a pursuit game played on graphs for which a minor is excluded, *J. Combin. Theory* (Ser. B) **41**, 37–47.

13. R.P. Anstee and M. Farber [1988], On bridged graphs and cop-win graphs, *J. Combin. Theory* (Ser. B) **44**, 22–28.

14. M. Ascher [1987], Mu Torere: An analysis of a Maori game, *Math. Mag.* **60**, 90–100.

15. I.M. Asel'derova [1974], On a certain discrete pursuit game on graphs, *Cybernetics* **10**, 859–864 (trans. of *Kibernetika* **10** (1974) 102–105).

16. R. Austin [1976], Impartial and partisan games, M.Sc. Thesis, Univ. of Calgary.

17. L. Babai and S. Moran [1988], Arthur-Merlin games: a randomized proof system, and a hierarchy of complexity classes, *J. Comput. System Sci.* **36**, 254–276.

18. W.W.R. Ball and H.S.M. Coxeter [1987], *Mathematical Recreations and Essays*, 13th ed., Dover, New York, NY.

19. R.B. Banerji [1971], Similarities in games and their use in strategy construction, Proc. Symp. Computers and Automata (J. Fox, ed.), Polytechnic Press, Brooklyn, NY, pp. 337–357.

20. R.B. Banerji [1980], *Artificial Intelligence, A Theoretical Approach*, Elsevier, North-Holland, New York, NY.

21. R.B. Banerji and G.W. Ernst [1972], Strategy construction using homomorphisms between games, *Artificial Intelligence* **3**, 223–249.

22. I. Bárány [1979], On a class of balancing games, *J. Combin. Theory* (Ser. A) **26**, 115–126.

23. J.G. Baron [1974], The game of nim — a heuristic approach, *Math. Mag.* **74**, 23–28.

24. J. Baumgartner, F. Galvin, R. Laver and R. McKenzie [1975], Game theoretic versions of partition relations, Colloquia Mathematica Societatis János Bolyai **10**, Proc. Internat. Colloq. on Infinite and Finite Sets, Vol. 1 (A. Hajnal, R. Rado and V.T. Sós, eds.), Keszthely, Hungary, 1973, North-Holland, pp. 131–135.

25. A. Beck [1969], Games, Chapter 5 in: A. Beck, M.N. Bleicher and D.W. Crowe, *Excursions into Mathematics*, Worth Publ., pp. 317–387.

26. J. Beck [1981], On positional games, *J. Combin. Theory* (Ser. A) **30**, 117–133.

27. J. Beck [1981], Van der Waerden and Ramsey type games, *Combinatorica* **1**, 103–116.

28. J. Beck [1982], On a generalization of Kaplansky's game, *Discrete Math.* **42**, 27–35.

29. J. Beck [1982], Remarks on positional games, I, *Acta Math. Acad. Sci. Hungar.* **40**(1–2), 65–71.

30. J. Beck [1985], Random graphs and positional games on the complete graph, *Ann. Discrete Math.* **28**, 7–13.

31. J. Beck and L. Csirmaz [1982], Variations on a game, *J. Combin. Theory* (Ser. A) **33**, 297–315.

32. R.C. Bell [1960, 1969], *Board and Table Games* (two volumes), Oxford University Press.

33. D. Berengut [1981], A random hopscotch problem or how to make Johnny read more, in: *The Mathematical Gardner* (D.A. Klarner, ed.), Wadsworth Internat., Belmont, CA, pp. 51–59.

34. C. Berge [1976], Sur lex jeux positionnels, *Cahiers du Centre Études Rech. Opér.* **18**, 91–107.

35. C. Berge [1977], Vers une théorie générale des jeux positionnels, Mathematical Economics and Game Theory, *Essays in honor of Oskar Morgentern*, (R. Henn and O. Moeschlin, eds.), *Lecture Notes in Economics* **141**, 13–24, Springer Verlag, Berlin.

36. C. Berge [1981], Some remarks about a Hex problem, in: *The Mathematical Gardner* (D.A. Klarner, ed.), Wadsworth Internat., Belmont, CA, pp. 25–27.

37. C. Berge [1985], *Graphs*, North-Holland, Amsterdam (Ch. 14).

38. C. Berge [1989], *Hypergraphs*, Elsevier (French: Gauthier Villars 1988) (Ch. 4).

39. C. Berge and M. Las Vergnas [1976], Un nouveau jeu positionnel, le "Match-It", ou une construction dialectique des couplages parfaits, *Cahiers du Centre Études Rech. Opér.* **18**, 83–89.

40. E.R. Berlekamp [1972], Some recent results on the combinatorial game called Welter's Nim, Proc. 6th Annual Princeton Conf. Information Science and Systems, pp. 203–204.

41. E.R. Berlekamp [1974], The Hackenbush number system for compresssion of numerical data, *Inform. and Control* **26**, 134–140.

42. E.R. Berlekamp [1988], Blockbusting and domineering, *J. Combin. Theory* (Ser. A) **49**, 67–116. An earlier version, entitled Introduction to blockbusting and domineering is to appear in: The Far Side of Mathematics, Proc. E. Strens Memorial Conf. on Recr. Math. and its History, Calgary, 1986, Math. Assoc. of America, Washington, D.C.

43. E.R. Berlekamp [1991], Introductory overview of mathematical Go endgames, in: Combinatorial Games, *Proc. Sympos. Appl. Math.* **43**, (R.K. Guy, ed.), Amer. Math. Soc., Providence, RI, pp. 73–100.

44. E.R. Berlekamp, J.H. Conway and R.K. Guy [1982], *Winning Ways* (two volumes), Academic Press, London.

45. P. Berloquin [1976], *100 Jeux de Table*, Flammarion, Paris.

46. D. Blackwell and M.A. Girshick [1954], *Theory of Games and Statistical Decisions*, Wiley, New York.

47. U. Blass and A.S. Fraenkel [1990], The Sprague-Grundy function of Wythoff's game, *Theoret. Comput. Sci.*, **75**, 311–333.

48. K.D. Boklan [1984], The n-number game, *Fibonacci Quart.* **22**, 152–155.

49. E. Borel [1921], La théorie du jeu et les équations integrales à noyau symmetrique gauche, *C.R. Acad. Sci. Paris* **173**, 1304–1308.

50. C.L. Bouton [1902], Nim, a game with a complete mathematical theory, *Ann. of Math.* **3**(2), 35–39.

51. J. Boyce [1981], A Kriegspiel endgame, in: *The Mathematical Gardner* (D.A. Klarner, ed.), Wadsworth Internat., Belmont, CA, pp. 28–36.

52. A.K. Chandra, D.C. Kozen and L.J. Stockmeyer [1981], Alternation, *J. Assoc. Comput. Mach.* **28**, 114–133.

53. A.K. Chandra and L.J. Stockmeyer [1976], Alternation, Proc. 17th Annual Symp. Foundations of Computer Science (Houston, TX, Oct. 1976), IEEE Computer Soc., Long Beach, CA, pp. 98–108.

54. S.M. Chase [1972], An implemented graph algorithm for winning Shannon switching games, *Commun. Assoc. Comput. Mach.* **15**, 253–256.

55. B.S. Chlebus [1986], Domino-tiling games, *J. Comput. System Sci.* **32**, 374–392.

56. F.R.K. Chung [1989], Pebbling in hypercubes, *SIAM J. Disc. Math.* **2**, 467–472.

57. F.R.K. Chung, J.E. Cohen and R.L. Graham [1988], Pursuit-evasion games on graphs, *J. Graph Theory* **12**, 159–167.

58. V. Chvátal [1973], On the computational complexity of finding a kernel, Report No. CRM-300, Centre de Recherches Mathématiques, Université de Montréal.

59. V. Chvátal [1981], Cheap, middling or dear, in: *The Mathematical Gardner* (D.A. Klarner, ed.), Wadsworth Internat., Belmont, CA, pp. 44–50.

60. V. Chvátal [1983], Mastermind, *Combinatorica* **3**, 325–329.

61. V. Chvátal and P. Erdős [1978], Biased positional games, *Ann. Discrete Math.* **2**: Algorithmic Aspects of Combinatorics, (B. Alspach, P. Hell and D.J. Miller, eds.), Qualicum Beach, B.C., Canada, 1976, North-Holland, pp. 221–229.

62. D.S. Clark [1986], Fibonacci numbers as expected values in a game of chance, *Fibonacci Quart.* **24**, 263–267.

63. A.J. Cole and A.J.T. Davie [1969], A game based on the Euclidean algorithm and a winning strategy for it, *Math. Gaz.* **53**, 354–357.

64. A. Condon and R.E. Ladner [1988], Probabilistic game automata, *J. Comput. System Sci.* **36**, 452–489.

65. I.G. Connell [1959], A generalization of Wythoff's game, *Canad. Math. Bull.* **2**, 181–190.

66. J.H. Conway [1972], All numbers great and small, Univ. of Calgary Math. Dept. Res. Paper No. 149.

67. J.H. Conway [1976], *On Numbers and Games*, Academic Press, London.

68. J.H. Conway [1977], All games bright and beautiful, *Amer. Math. Monthly* **84**, 417–434.

69. J.H. Conway [1978], A gamut of game theories, *Math. Mag.* **51**, 5–12.

70. J.H. Conway [1978], Loopy Games, *Ann. Discrete Math.* **3**: Proc. Symp. Advances in Graph Theory, Cambridge Combinatorial Conf. (B. Bollobás, ed.), Cambridge, May 1977, pp. 55–74.

71. J.H. Conway [1990], Integral lexicographic codes, *Discrete Math.*, **83**, 219–235.

72. J.H. Conway [1991], More ways of combining games, in: Combinatorial Games, *Proc. Sympos. Appl. Math.* **43**, (R.K. Guy, ed.), Amer. Math. Soc., Providence, RI, pp. 57–72.

73. J.H. Conway [1991], Numbers and games, in: Combinatorial Games, *Proc. Symp. Appl. Math.* **43**, (R.K. Guy, ed.), Amer. Math. Soc., Providence, RI, pp. 23–34.

74. J.H. Conway and H.S.M. Coxeter [1973], Triangulated polygons and frieze patterns, *Math. Gaz.* **57**, 87–94; 175–183.

75. J.H. Conway and N.J.A. Sloane [1986], Lexicographic codes: error-correcting codes from game theory, *IEEE Trans. Inform. Theory* **IT-32**, 337–348.

76. H.S.M. Coxeter [1953], The golden section, phyllotaxis and Wythoff's game, *Scripta Math.* **19**, 135–143.

77. J.W. Creely [1987], The length of a two-number game, *Fibonacci Quart.* **25**, 174–179.

78. J.W. Creely [1988], The length of a three-number game, *Fibonacci Quart.* **26**, 141–143.

79. L. Csirmaz [1980], On a combinatorial game with an application to Go-Moku, *Discrete Math.* **29**, 19–23.

80. L. Csirmaz and Zs. Nagy [1979], On a generalization of the game Go-Moku, II, *Studia Scientiarum Math. Hung.* **14**, 461–469.

81. J. Czyzowicz, D. Mundici and A. Pelc [1988], Solution of Ulam's problem on binary search with two lies, *J. Combin. Theory* (Ser. A) **49**, 384–388.

82. J. Czyzowicz, D. Mundici and A. Pelc [1989], Ulam's searching game with lies, *J. Combin. Theory* (Ser. A) **52**, 62–76.

83. G. Danaraj and V. Klee [1977], The connectedness game and the c-complexity of certain graphs, *SIAM J. Appl. Math.* **32**, 431–442.

84. M. Davis [1963], Infinite games of perfect information, *Ann. of Math. Stud.* **52**, 85–101, Princeton.

85. T.R. Dawson [1934], *Fairy Chess Review* (problem 1603, p. 94, Dec.).

86. T.R. Dawson [1935], Caissa's Wild Roses, Reprinted in: *Five Classics of Fairy Chess*, Dover, 1973.

87. N.G. de Bruijn [1981], Pretzel Solitaire as a pastime for the lonely mathematician, in: *The Mathematical Gardner* (D.A. Klarner, ed.), Wadsworth Internat., Belmont, CA, pp. 16–24.

88. F. de Carteblanche [1970], The princess and the roses, *J. Recr. Math.* **3**, 238–239.

89. F. deCarte Blanche [1974], The roses and the princes, *J. Recr. Math.* **7**, 295–298.

90. B. Descartes [1953], Why are series musical? *Eureka* **16**, 18–20. Reprinted *ibid.* **27** (1964) 29–31.

91. A.K. Dewdney, Computer Recreations, a column in Scientific American (since May, 1984).

92. A.P. Domoryad [1964], *Mathematical Games and Pastimes* (translated by H. Moss), Pergamon Press, Oxford.

93. P. Duchet [1987], A sufficient condition for a digraph to be kernel-perfect, *J. Graph Theory* **11**, 81–85.

94. P. Duchet [1987], Parity graphs are kernel-M-solvable, *J. Combin. Theory* (Ser. B) **43**, 121–126.

95. H.E. Dudeney [1958], *The Canterbury Puzzles*, reprinted by Dover, New York, NY.

96. H.E. Dudeney [1989], *Amusements in Mathematics*, reprinted by Dover, New York, NY.

97. N. Duvdevani and A.S. Fraenkel [1989], Properties of k-Welter's game, *Discrete Math.* **76**, 197–221.

98. J. Edmonds [1965], Lehman's switching game and a theorem of Tutte and Nash-Williams, *J. Res. Nat. Bur. Standards* **69B**, 73–77.

99. A. Ehrenfeucht and J. Mycielski [1979], Positional strategies for mean payoff games, *Internat. J. Game Theory* **8**, 109–113.

100. R.J. Epp and T.S. Ferguson [1980], A note on take-away games, *Fibonacci Quart.* **18**, 300–303.

101. R.A. Epstein [1977], *Theory of Gambling and Statistial Logic*, Academic Press, New York, NY.

102. P. Erdős and J.L. Selfridge [1973], On a combinatorial game, *J. Combin. Theory* (Ser. A) **14**, 298–301.

103. M. Erickson and F. Harary [1983], Picasso animal achievement games, *Bull. Malaysian Math. Soc.* **6**, 37–44.

104. R.J. Evans [1974], A winning opening in reverse Hex, *J. Recr. Math.* **7**, 189–192.

105. R.J. Evans [1975–76], Some variants of Hex, *J. Recr. Math.* **8**, 120–122.

106. R.J. Evans [1979], Silverman's game on intervals, *Amer. Math. Monthly* **86**, 277–281.

107. S. Even and R.E. Tarjan [1976], A combinatorial problem which is complete in polynomial space, *J. Assoc. Comput. Mach.* **23**, 710–719. Also appeared in 7th Annual ACM Symp. Theory of Computing (Albuquerque, NM, 1975), Assoc. Comput. Mach., New York, NY, 1975, pp. 66–71.

108. E. Falkener [1961], *Games Ancient and Modern*, Dover, New York, NY. (Published previously by Longmans Green, 1897.)

109. T.S. Ferguson [1974], On sums of graph games with last player losing, *Internat. J. Game Theory* **3**, 159–167.

110. T.S. Ferguson [1984], Misère annihilation games, *J. Combin. Theory* (Ser. A) **37**, 205–230.

111. A.S. Finbow and B.L. Hartnell [1983], A game related to covering by stars, *Ars Combinatoria* **16-A**, 189–198.

112. P.C. Fishburn and N.J.A. Sloane [1989], The solution to Berlekamp's switching game, *Discrete Math.* **74**, 263–290.

113. J.A. Flanigan [1978], Generalized two-pile Fibonacci nim, *Fibonacci Quart.* **16**, 459–469.

114. J.A. Flanigan [1981], On the distribution of winning moves in random game trees, *Bull. Austr. Math. Soc.* **24**, 227–237.

115. J.A. Flanigan [1981], Selective sums of loopy partizan graph games, *Internat. J. Game Theory* **10**, 1–10.

116. J.A. Flanigan [1982], A complete analysis of black-white Hackendot, *Internat. J. Game Theory* **11**, 21–25.

117. J.A. Flanigan [1982], One-pile time and size dependent take-away games, *Fibonacci Quart.* **20**, 51–59.

118. J.A. Flanigan [1983], Slow joins of loopy games, *J. Combin. Theory* (Ser. A) **34**, 46–59.

119. A.S. Fraenkel [1974], Combinatorial games with an annihilation rule, in: The Influence of Computing on Mathematical Research and Education, *Proc. Sympos. Appl. Math.* (J.P. LaSalle, ed.), Vol. 20, Amer. Math. Soc., Providence, RI, pp. 87–91.

120. A.S. Fraenkel [1977], The particles and antiparticles game, *Comput. Math. Appl.* **3**, 327–328.

121. A.S. Fraenkel [1980], From Nim to Go, *Ann. Discrete Math.* **6**: Proc. Symp. on Combinatorial Mathematics, Combinatorial Designs and Their Applications (J. Srivastava, ed.), Colorado State Univ., Fort Collins, CO, June 1978, pp. 137–156.

122. A.S. Fraenkel [1981], Planar kernel and Grundy with $d \leq 3$, $d_{out} \leq 2$, $d_{in} \leq 2$ are NP-complete, *Discrete Appl. Math.* **3**, 257–262.

123. A.S. Fraenkel [1982], How to beat your Wythoff games' opponents on three fronts, *Amer. Math. Monthly* **89**, 353–361.

124. A.S. Fraenkel [1983], 15 Research problems on games, in: "Research Problems", *Discrete Math.* **43–46**.

125. A.S. Fraenkel [1984], Wythoff games, continued fractions, cedar trees and Fibonacci searches, *Theoret. Comput. Sci.* **29**, 49–73. An earlier version appeared in Proc. 10th Internat. Colloq. on Automata, Languages and Programming (J. Diaz, ed.), Barcelona, July 1983, *Lecture Notes in Computer Science* **154**, 203–225, Springer Verlag, Berlin, 1983.

126. A.S. Fraenkel [1991], Recreation and depth in combinatorial games, in: The Far Side of Mathematics, Proc. E. Strens Memorial Conf. on Recr. Math. and its History, Calgary, 1986, Math. Assoc. of America, Washington, D.C., in press.

127. A.S. Fraenkel [1991], Complexity of games, in: Combinatorial Games, *Proc. Sympos. Appl. Math.* **43**, (R.K. Guy, ed.), Amer. Math. Soc., Providence, RI, pp. 111–154.

128. A.S. Fraenkel and I. Borosh [1973], A generalization of Wythoff's game, *J. Combin. Theory* (Ser. A) **15**, 175–191.

129. A.S. Fraenkel, M.R. Garey, D.S. Johnson, T. Schaefer and Y. Yesha [1978], The complexity of checkers on an $n \times n$ board — preliminary report, Proc. 19th Annual Symp. Foundations of Computer Science (Ann Arbor, MI, Oct. 1978), IEEE Computer Soc., Long Beach, CA, pp. 55–64.

130. A.S. Fraenkel and E. Goldschmidt [1987], Pspace-hardness of some combinatorial games, *J. Combin. Theory* (Ser. A) **46**, 21–38.

131. A.S. Fraenkel and F. Harary [1989], Geodetic contraction games on graphs, *Internat. J. Game Theory* **18**, 327–338.

132. A.S. Fraenkel and H. Herda [1980], Never rush to be first in playing Nimbi, *Math. Mag.* **53**, 21–26.

133. A.S. Fraenkel and A. Kotzig [1987], Partizan octal games: partizan subtraction games, *Internat. J. Game Theory* **16**, 145–154.

134. A.S. Fraenkel and D. Lichtenstein [1981], Computing a perfect strategy for $n \times n$ chess requires time exponential in n, *J. Combin. Theory* (Ser. A) **31**, 199–214. Preliminary version in Proc. 8th Internat. Colloq. Automata, Languages and Programming (S. Even and O. Kariv, eds.), Acre, Israel, 1981, *Lecture Notes in Computer Science* **115**, 278–293, Springer Verlag, Berlin.

135. A.S. Fraenkel, M. Loebl and J. Nešetřil [1988], Epidemiography, II. Games with a dozing yet winning player, *J. Combin. Theory* (Ser. A) **49**, 129–144.

136. A.S. Fraenkel and M. Lorberbom [1989], Epidemiography with various growth functions, *Discrete Appl. Math.* **25**, 53–71.

137. A.S. Fraenkel and M. Lorberbom [≥1990], Nimhoff games, *J. Combin. Theory* (Ser. A), in press.

138. A.S. Fraenkel and J. Nešetřil [1985], Epidemiography, *Pacific J. Math.* **118**, 369–381.

139. A.S. Fraenkel and Y. Perl [1975], Constructions in combinatorial games with cycles, *Coll. Math. Soc. János Bolyai* **10**: Proc. Internat. Colloq. on Infinite and Finite Sets, Vol. 2 (A. Hajnal, R. Rado and V.T. Sós, eds.) Keszthely, Hungary, 1973, North-Holland, pp. 667–699.

140. A.S. Fraenkel and E.R. Scheinerman [≥1990], A deletion game on hypergraphs, *Discrete Appl. Math.*, in press.

141. A.S. Fraenkel and U. Tassa [1975], Strategy for a class of games with dynamic ties, *Comput. Math. Appl.* **1**, 237–254.

142. A.S. Fraenkel and U. Tassa [1982], Strategies for compounds of partizan games, *Math. Proc. Camb. Phil. Soc.* **92**, 193–204.

143. A.S. Fraenkel, U. Tassa and Y. Yesha [1978], Three annihilation games, *Math. Mag.* **51**, 13–17.

144. A.S. Fraenkel and Y. Yesha [1976], Theory of annihilation games, *Bull. Amer. Math. Soc.* **82**, 775–777.

145. A.S. Fraenkel and Y. Yesha [1979], Complexity of problems in games, graphs and algebraic equations, *Discrete Appl. Math.* **1**, 15–30.

146. A.S. Fraenkel and Y. Yesha [1982], Theory of annihilation games — I, *J. Combin. Theory* (Ser. B) **33**, 60–86.

147. A.S. Fraenkel and Y. Yesha [1986], The generalized Sprague-Grundy function and its invariance under certain mappings, *J. Combin. Theory* (Ser. A) **43**, 165–177.

148. P. Frankl [1987], Cops and robbers in graphs with large girth and Cayley graphs, *Discrete Appl. Math.* **17**, 301–305.

149. P. Frankl [1987], On a pursuit game on Cayley graphs, *Combinatorica* **7**, 67–70.

150. D. Fremlin [1973], Well-founded games, *Eureka* **36**, 33–37.

151. D. Gale [1974], A curious Nim-type game, *Amer. Math. Monthly* **81**, 876–879.

152. D. Gale [1979], The game of Hex and the Brouwer fixed-point theorem, *Amer. Math. Monthly* **86**, 818–827.

153. D. Gale and A. Neyman [1982], Nim-type games, *Internat. J. Game Theory* **11**, 17–20.

154. D. Gale and F.M. Stewart [1953], Infinite games with perfect information, Contributions to the Theory of Games, *Ann. of Math. Stud.* **2**(28), 245–266, Princeton.

155. F. Galvin [1978], Indeterminacy of point-open games. *Bull. de l'Academie Polonaise des Sciences* (math., astr. et phys.) **26**, 445–449.

156. A. Gangolli and T. Plambeck [1989], A note on periodicity in some octal games, *Internat. J. Game Theory* **18**, 311–320.

157. M. Gardner [1956], *Mathematics, Magic and Mystery*, Dover, New York, NY.

158. M. Gardner, Mathematical Games, a column in Scientific American (Jan. 1957–Dec. 1981).

159. M. Gardner [1959], *The Scientific American Book of Mathematical Puzzles and Diversions*, Simon & Schuster, NY.

160. M. Gardner [1959], *Mathematical Puzzles of Sam Loyd*, Dover, New York, NY.

161. M. Gardner [1960], *More Mathematical Puzzles of Sam Loyd*, Dover, New York, NY.

162. M. Gardner [1961], *The Second Scientific American Book of Mathematical Puzzles and Diversions*, Simon and Schuster, NY.

163. M. Gardner [1966], *New Mathematical Diversions from Scientific American*, Simon and Schuster, New York, NY.

164. M. Gardner [1967], *536 Puzzles and Curious Problems* (edited; reissue of H.E. Dudeney's *Modern Puzzles* and *Puzzles and Curious Problems*), Scribner's, NY.

165. M. Gardner [1969], *The Unexpected Hanging and Other Mathematical Diversions*, Simon and Schuster, NY.

166. M. Gardner [1975], *Mathematical Carnival*, Knopf, NY.

167. M. Gardner [1977], *Mathematical Magic Show*, Knopf, NY.

168. M. Gardner [1978], *Aha! Insight*, Freeman, New York, NY.

169. M. Gardner [1979], *Mathematical Circus*, Knopf, NY.

170. M. Gardner [1981], *Science Fiction Puzzle Tales*, Potter.

171. M. Gardner [1982], *Aha! Gotcha!*, Freeman, New York, NY.

172. M. Gardner [1984], *Puzzles from Other Worlds*, Random House.

173. M. Gardner [1984], *The Magic Numbers of Dr. Matrix*, Prometheus.

174. M. Gardner [1984], *The Sixth Book of Mathematical Games*, Univ. of Chicago Press (first appeared in 1971 by Freeman).

175. M. Gardner [1984], *Wheels, Life and Other Mathematical Amusements*, Freeman, New York, NY.

176. M. Gardner [1986], *Knotted Doughnuts and Other Mathematical Entertainments*, Freeman, New York, NY.

177. M. Gardner [1987], *Riddles of the Sphinx*, Math. Assoc. of America.

178. M. Gardner [1987], *Time Travel and Other Mathematical Bewilderments*, Freeman, New York, NY.

179. M. Gardner [1989], *Penrose Tiles to Trapdoor Ciphers*, Freeman, New York, NY.

180. M.R. Garey and D.S. Johnson [1979], *Computers and Intractability: A Guide to the theory of NP-Completeness*, Freeman, San Francisco (Appendix, A8).

181. J.R. Gilbert, T. Lengauer and R.E. Tarjan [1980], The pebbling problem is complete in polynomial space, *SIAM J. Comput.* **9**, 513–524. Preliminary version in Proc. 11th Annual ACM Symp. Theory of Computing (Atlanta, GA, 1979), Assoc. Comput. Mach., New York, NY, pp. 237–248.

182. S.W. Golomb [1965], *Polyominoes*, Allen and Unwin, London.

183. S.W. Golomb [1966], A mathematical investigation of games of "take-away", *J. Combin. Theory* **1**, 443–458.

184. S.B. Grantham [1985], Galvin's "racing pawns" game and a well-ordering of trees, *Memoirs Amer. Math. Soc* **53** (316), 63 pp.

185. P.M. Grundy [1939], Mathematics and Games, *Eureka* **2**, 6–8; reprinted in *Eureka* **27** (1964), 9–11.

186. P.M. Grundy, R.S. Scorer and C.A.B. Smith [1944], Some binary games, *Math. Gaz.* **28**, 96–103.

187. P.M. Grundy and C.A.B. Smith [1956], Disjunctive games with the last player losing, *Proc. Camb. Phil. Soc.* **52**, 527–533.

188. F. Grunfeld and R.C. Bell [1975], *Games of the World*, Holt, Rinehart and Winston.

189. C.D. Grupp [1976], *Brettspiele-Denkspiele*, Humboldt-Taschenbuchverlag, München.

190. R.K. Guy [1976], Packing $[1, n]$ with solutions of $ax + by = cz$; the unity of combinatorics, *Atti Conv. Lincei* #17, *Accad. Naz. Lincei* Rome, Tomo II, 173–179.

191. R.K. Guy [1976], Twenty questions concerning Conway's Sylver Coinage, *Amer. Math. Monthly* **83**, 634–637.

192. R.K. Guy [1977], Games are graphs, indeed they are trees, *Proc. 2nd Carib. Conf. Combin. and Comput.*, Letchworth Press, Barbados, 6–18.

193. R.K. Guy [1977], She loves me, she loves me not; relatives of two games of Lenstra, *Een Pak med een Korte Broek* (papers presented to H.W. Lenstra), Mathematisch Centrum, Amsterdam.

194. R.K. Guy [1978], Partizan and impartial combinatorial games, *Colloq. Math. Soc. János Bolyai* **18**: Proc. 5th Hungar. Conf. Combin., Vol. I (A. Hajnal and V.T. Sós, eds.), Keszthely, Hungary, 1976, North-Holland, pp. 437–461.

195. R.K. Guy [1979], Partizan Games, Colloques Internationaux C.N.R. No. 260 — Problèmes Combinatoires et Théorie des Graphs, 199–205.

196. R.K. Guy [1981], Anyone for twopins?, in: *The Mathematical Gardner* (D.A. Klarner, ed.), Wadsworth Internat., Belmont, CA, pp. 2–15.

197. R.K. Guy [1983], Graphs and games, in: *Selected Topics in Graph Theory*, 2 (L.W. Beineke and R.J. Wilson, eds.), Academic Press, London, pp. 269–295.

198. R.K. Guy [1986], John Isbell's game of beanstalk and John Conway's game of beans-don't-talk, *Math. Mag.* **59**, 259–269.

199. R.K. Guy [1989], *Fair Game*, COMAP Math. Exploration Series, Arlington, MA.

200. R.K. Guy [≥1990], Combinatorial games, in: *Handbook of Combinatorics* (Graham, Grötschel and Lovász, eds.) North-Holland, Amsterdam, in press.

201. R.K. Guy [1991], Impartial Games, in: Combinatorial Games, *Proc. Sympos. Appl. Math.* **43**, (R.K. Guy, ed.), Amer. Math. Soc., Providence, RI, pp. 35–56.

202. R.K. Guy [1991], What is a game? in: Combinatorial Games, *Proc. Sympos. Appl. Math.* **43**, (R.K. Guy, ed.), Amer. Math. Soc., Providence, RI, pp. 1–22.

203. R.K. Guy [ ], Mathematics from fun & fun from mathematics; an informal autobiographical history of combinatorial games, submitted.

204. R.K. Guy and C.A.B. Smith [1956], The G-values of various games, *Proc. Camb. Phil. Soc.* **52**, 514–526.

205. W. Guzicki [1990], Ulam's searching game with two lies, *J. Combin. Theory* (Ser. A) **54**, 1–19.

206. D.R. Hale [1983], A variant of Nim and a function defined by Fibonacci representation, *Fibonacci Quart.* **21**, 139–142.

207. A.W. Hales and R.I. Jewett [1963], Regularity and positional games, *Trans. Amer. Math. Soc.* **106**, 222–229.

208. Y.O. Hamidoune [1987], On a pursuit game of Cayley digraphs, *Europ. J. Combin.* **8**, 289–295.

209. Y.O. Hamidoune and M. Las Vergnas [1985], The directed Shannon switching game and the one-way game, in: *Graph Theory and Its Applications to Algorithms and Computer Science* (Y. Alavi et al., eds.), Wiley, pp. 391–400.

210. Y.O. Hamidoune and M. Las Vergnas [1986], Directed switching games on graphs and matroids, *J. Combin. Theory* (Ser. B) **40**, 237–269.

211. Y.O. Hamidoune and M. Las Vergnas [1987], A solution to the box game, *Discrete Math.* **65**, 157–171.

212. Y.O. Hamidoune and M. Las Vergnas [1988], A solution to the misère Shannon switching game, *Discrete Math.* **72**, 163–166.

213. O. Hanner [1959], Mean play of sums of positional games, *Pacific J. Math.* **9**, 81–99.

214. F. Harary [1982], Achievement and avoidance games for graphs, *Ann. Discrete Math.* **13**, 111–120.

215. P. Hein [1942], Polygon, *Politiken* (description of Hex in this Danish newspaper of Dec. 26).

216. D. Hensley [1988], A winning strategy at Taxman, *Fibonacci Quart.* **26**, 262–270.

217. G.A. Heuer [1982], Odds versus evens in Silverman-type games, *Internat. J. Game Theory* **11**, 183–194.

218. G.A. Heuer [1989], Reduction of Silverman-like games to games on bounded sets, *Internat. J. Game Theory* **18**, 31–36.

219. G.A. Heuer and W.D. Rieder [1988], Silverman games on disjoint discrete sets, *SIAM J. Disc. Math.* **1**, 485–525.

220. P.G. Hinman [1972], Finite termination games with tie, *Israel J. Math.* **12**, 17–22.

221. S. Hitotumatu [1968], Some remarks on nim-like games, *Comment. Math. Univ. St. Paul* **17**, 85–98.

222. J.C. Holladay [1957], Cartesian products of termination games, Contributions to the Theory of Games, *Ann. of Math. Stud.* **3**(39), 189–200, Princeton.

223. J.C. Holladay [1958], Matrix nim, *Amer. Math. Monthly* **65**, 107–109.

224. J.C. Holladay [1966], A note on the game of dots, *Amer. Math. Monthly* **73**, 717–720.

225. O. Itzinger [1977], The South American game, *Math. Mag.* **50**, 17–21.

226. T.A. Jenkyns and J.P. Mayberry [1980], The skeleton of an impartial game and the Nim-function of Moore's Nim_k, *Internat. J. Game Theory* **9**, 51–63.

227. D.S. Johnson [1983], The NP-Completeness Column: An Ongoing Guide, *J. Algorithms* **4**, 397–411 (9th quarterly column (games); column started in 1981).

228. J.P. Jones [1982], Some undecidable determined games, *Internat. J. Game Theory* **11**, 63–70.

229. M. Kac [1974], Hugo Steinhaus, a reminiscence and a tribute, *Amer. Math. Monthly* **81**, 572–581 (p. 577).

230. J. Kahane and A.S. Fraenkel [1987], k-Welter — a generalization of Welter's game, *J. Combin. Theory* (Ser. A) **46**, 1–20.

231. J. Kahn, J.C. Lagarias and H.S. Witsenhausen [1987], Single-suit two-person card play, *Internat. J. Game Theory* **16**, 291–320.

232. J. Kahn, J.C. Lagarias and H.S. Witsenhausen [1988], Single-suit two-person card play, II. Dominance, *Order* **5**, 45–60.

233. J. Kahn, J.C. Lagarias and H.S. Witsenhausen [1989], Single-suit two-person card play, III. The misère game, *SIAM J. Disc. Math.* **2**, 329–343.

234. J. Kahn, J.C. Lagarias and H.S. Witsenhausen [1989], On Lasker's card game, in: Differential games and applications (T.S. Başar, P. Bernhard, eds.), *Lecture Notes in Control and Information Sciences* **119**, 1–8, Springer Verlag, Berlin.

235. L. Kalmár [1928], Zur Theorie der abstrakten Spiele, *Acta Sci. Math. Univ. Szeged* **4**, 65–85.

236. B. Kalyanasundaram and G. Schnitger [1988], On the power of white pebbles, Proc. 20th Annual ACM Symp. Theory of Computing (Chicago, IL, 1988), Assoc. Comput. Mach., New York, NY, pp. 258–266.

237. R.M. Karp and Y. Zhang [1989], On parallel evaluation of game trees, Proc. ACM Symp. Parallel Algorithms and Architectures, 409–420.

238. T. Kasai, A. Adachi and S. Iwata [1979], Classes of pebble games and complete problems, *SIAM J. Comput.* **8**, 574–586.

239. J.C. Kenyon [1967], A Nim-like game with period 349, Univ. of Calgary, Math. Dept. Res. Paper No. 13.

240. J.C. Kenyon [1967], Nim-like games and the Sprague-Grundy theory, M.Sc. Thesis, Univ. of Calgary.

241. B. Kerst [1933], *Mathematische Spiele*, Reprinted by Dr. Martin Sändig oHG, Wiesbaden 1968.

242. H. Kinder [1973], Gewinnstrategien des Anziehenden in einigen Spielen auf Graphen, *Arch. Math.* **24**, 332–336.

243. M.M. Klawe [1985], A tight bound for black and white pebbles on the pyramids, *J. Assoc. Comput. Mach.* **32**, 218–228.

244. D.J. Kleitman and B.L. Rothschild [1972], A generalization of Kaplansky's game, *Discrete Math.* **2**, 173–178.

245. D.E. Knuth [1974], *Surreal Numbers*, Addison-Wesley, Reading, MA.

246. D.E. Knuth [1976], The computer as Master Mind, *J. Recr. Math.* **9**, 1–6.

247. A. Kotzig [1946], O k-posunutiach (On k-translations; in Slovakian), *Časop. pro Pěst. Mat. a Fys.* **71**, 55–61; Extended abstract in French, pp. 62–66.

248. G. Kowalewski [1930], *Alte und neue mathematische Spiele*, Reprinted by Dr. Martin Sändig oHG, Wiesbaden 1968.

249. B. Kummer [1980], *Spiele auf Graphen*, Internat. Series of Numerical Mathematics, Birkhäuser Verlag, Basel.

250. R.E. Ladner and J.K. Norman [1985], Solitaire automata, *J. Comput. System Sci.* **30**, 116–129.

251. J.C. Lagarias [1977], Discrete balancing games, *Bull. Inst. Math. Acad. Sinica* **5**, 363-373.

252. E. Lasker [1931], *Brettspiele der Völker, Rätsel und mathematische Spiele*, Berlin.

253. A. Lehman [1964], A solution to the Shannon switching game, *SIAM J. Appl. Math.* **12**, 687–725.

254. T. Lengauer and R. Tarjan [1980], The space complexity of pebble games on trees, *Inform. Process. Lett.* **10**, 184–188.

255. T. Lengauer and R. Tarjan [1982], Asymptotically tight bounds on time-space trade-offs in a pebble game, *J. Assoc. Comput. Mach.* **29**, 1087–1130.

256. H.W. Lenstra, Jr. [1977], On the algebraic closure of two, *Proc. Kon. Nederl. Akad. Wetensch.* (Ser. A) **80**, 389–396.

257. H.W. Lenstra, Jr. [1977/1978], Nim multiplication, Séminaire de Théorie des Nombres, No. 11, Université de Bordeaux, France.

258. S.-Y.R. Li [1974], Generalized impartial games, *Internat. J. Game Theory* **3**, 169–184.

259. S.-Y.R. Li [1976], Sums of Zuchswang games, *J. Combin. Theory* (Ser. A) **21**, 52–67.

260. S.-Y.R. Li [1977], N-person nim and N-person Moore's games, *Internat. J. Game Theory* **7**, 31–36.

261. D. Lichtenstein, [1982], Planar formulae and their uses, *SIAM J. Comput.* **11**, 329–343.

262. D. Lichtenstein and M. Sipser [1980], Go is Polynomial-Space hard, *J. Assoc. Comput. Mach.* **27**, 393–401. Earlier draft appeared in Proc. 19th Annual Symp. Foundations of Computer Science (Ann Arbor, MI, Oct. 1978), IEEE Computer Soc., Long Beach, CA, 1978, pp. 48–54.

263. S. Loyd [1914], *Cyclopedia of Puzzles and Tricks*, Franklin Bigelow Corporation, Morningside Press, NY. Reissued and edited by M. Gardner under the name *The Mathematical Puzzles of Sam Loyd* (two volumes), Dover, New York, NY, 1959.

264. E. Lucas [1960], *Recreations Mathematiques* (four volumes), A. Blanchard, Paris.

265. A.L. Ludington [1988], Length of the 7-number game, *Fibonacci Quart.* **26**, 195–204.

266. M. Maamoun and H. Meyniel [1987], On a game of policemen and robber, *Discrete Appl. Math.* **17**, 307–309.

267. P.A. MacMahon [1921], *New Mathematical Pastimes*, Cambridge University Press, Cambridge.

268. J.G. Mauldon [1978], Num, a variant of nim with no first player win, *Amer. Math. Monthly* **85**, 575–578.

269. D.P. McIntyre [1942], A new system for playing the game of nim, *Amer. Math. Monthly* **49**, 44–46.

270. E. Mead, A. Rosa and C. Huang [1974], The game of SIM: A winning strategy for the second player, *Math. Mag.* **47**, 243–247.

271. N. Megiddo, S.L. Hakimi, M.R. Garey, D.S. Johnson and C.H. Papadimitriou [1988], The complexity of searching a graph, *J. Assoc. Comput. Mach.* **35**, 18–44.

272. K. Mehlhorn, S. Näher and M. Rauch [1989], On the complexity of a game related to the dictionary problem, Proc. 30th Annual Symp. Foundations of Computer Science, pp. 546–548.

273. C.G. Méndez [1981], On the law of large numbers, infinite games and category, *Amer. Math. Monthly* **88**, 40–42.

274. F. Meyer auf der Heide [1981], A comparison of two variations of a pebble game on graphs, *Theoret. Comput. Sci.* **13**, 315–322.

275. H. Meyniel and J.-P. Roudneff [1988], The vertex picking game and a variation of the game of dots and boxes, *Discrete Math.* **70**, 311–313.

276. J. Milnor [1953], Sums of positional games, Contributions to the Theory of Games, (H.W. Kuhn and A.W. Tucker, eds.), *Ann. of Math. Stud.* **2**(28), 291–301, Princeton.

277. P.-Min Lin [1982], Principal partition of graphs and connectivity games, *J. Franklin Inst.* **314**, 203–210.

278. E.H. Moore [1909–1910], A generalization of the game called nim, *Ann. of Math.* **11** (Ser. 2), 93–94.

279. F.L. Morris [1981], Playing disjunctive sums is polynomial space complete, *Internat. J. Game Theory* **10**, 195–205.

280. H.J.R. Murray [1952], *A History of Board Games Other Than Chess*, Oxford University Press.

281. R.J. Nowakowski [1991], ...,Welter's game, Sylver coinage, dots-and-boxes, ..., in: Combinatorial Games, *Proc. Sympos. Appl. Math.* **43**, (R.K. Guy, ed.), Amer. Math. Soc., Providence, RI, pp. 155-182.

282. R. Nowakowski and P. Winkler [1983], Vertex-to-vertex pursuit in a graph, *Discrete Math.* **43**, 235–239.

283. T.H. O'Beirne [1984], *Puzzles and Paradoxes*, Dover, New York, NY. (Appeared previously by Oxford University Press, London, 1965.)

284. E.W. Packel [1987], The algorithm designer versus nature: a game-theoretic approach to information-based complexity, *J. Complexity* **3**, 244–257.

285. C.H. Papadimitriou [1985], Games against nature, *J. Comput. System Sci.* **31**, 288–301.

286. A. Papaioannou [1982], A Hamiltonian game, *Ann. Discrete Math.* **13**, 171–178.

287. O. Patashnik [1980], Qubic: $4 \times 4 \times 4$ Tic-Tac-Toe, *Math. Mag.* **53**, 202–216.

288. J.L. Paul [1978], Tic-Tac-Toe in n dimensions, *Math. Mag.* **51**, 45–49.

289. W.J. Paul, E.J. Prauss and R. Reischuk [1980], On alternation, *Acta Informatica* **14**, 243–255.

290. W.J. Paul and R. Reischuk, On alternation, II [1980], *Acta Informatica* **14**, 391–403.

291. J. Pearl [1980], Asymptotic properties of minimax trees and game-searching procedures, *Artificial Intelligence* **14**, 113–138.

292. J. Pearl [1984], *Heuristics: Intelligent Search Strategies for Computer Problem Solving*, Addison-Wesley, Reading, MA.

293. A. Pelc [1987], Solution of Ulam's problem on searching with a lie, *J. Combin. Theory* (Ser. A) **44**, 129–140.

294. A. Pelc [1988], Prefix search with a lie, *J. Combin. Theory* (Ser. A) **48**, 165–173.

295. A. Pelc [1989], Detecting errors in searching games, *J. Combin. Theory* (Ser. A) **51**, 43–54.

296. D.H. Pelletier [1987], Merlin's magic square, *Amer. Math. Monthly* **94**, 143–150.

297. G.L. Peterson, Press-Ups is Pspace-complete, Unpublished manuscript.

298. G.L. Peterson and J.H. Reif [1979], Multiple-person alternation, Proc. 20th Annual Symp. Foundations Computer Science (San Juan, Puerto Rico, Oct. 1979), IEEE Computer Soc., Long Beach, CA, pp. 348–363.

299. N. Pippenger [1980], Pebbling, Proc. 5th IBM Symp. Math. Foundations of Computer Science, IBM, Japan (19 pp.)

300. N. Pippenger [1982], Advances in pebbling, in: Proc. 9th Internat. Colloq. Automata, Languages and Programming (M. Nielson and E.M. Schmidt, eds.), *Lecture Notes in Computer Science* **140**, 407–417, Springer Verlag, New York, NY.

301. V. Pless [1991], Games and codes, in: Combinatorial Games, *Proc. Sympos. Appl. Math.* **43**, (R.K. Guy, ed.), Amer. Math. Soc., Providence, RI, pp. 101–110.

302. J. Propp [≥1991], A new take-away game, in: The Far Side of Mathematics, Proc. E. Strens Memorial Conf. on Recr. Math and its History, Calgary, 1986, Math. Assoc. of America, Washington, D.C., in press.

303. A. Pultr and F.L. Morris [1984], Prohibiting repetitions makes playing games substantially harder, *Internat. J. Game Theory* **13**, 27–40.

304. A. Quilliot [1982], Discrete pursuit games, Proc. 13th Conference on Graphs and Combinatorics, Boca Raton, FL.

305. A. Quilliot [1985], A short note about pursuit games played on a graph with a given genus, *J. Combin. Theory* (Ser. B) **38**, 89–92.

306. M.O. Rabin [1957], Effective computability of winning strategies, Contributions to the Theory of Games, *Ann. of Math. Stud.* **3**(39), 147–157, Princeton.

307. M.O. Rabin [1976], Probabilistic algorithms, Proc. Symp. on New Directions and Recent Results in Algorithms and Complexity (J.F. Traub, ed.), Carnegie-Mellon, Academic Press, New York, NY, pp. 21–39.

308. B. Ravikumar and K.B. Lakshmanan [1984], Coping with known patterns of lies in a search game, *Theoret. Comput. Sci.* **33**, 85–94.

309. J.H. Reif [1984], The complexity of two-player games of incomplete information, *J. Comput. System Sci.* **29**, 274–301.

Earlier draft entitled Universal games of incomplete information, appeared in Proc. 11th Annual ACM Symp. Theory of Computing (Atlanta, GA, 1979), Assoc. Comput. Mach., New York, NY, pp. 288–308.

310. S. Reisch [1980], Gobang ist PSPACE-vollständig, *Acta Informatica* **13**, 59–66.

311. S. Reisch [1981], Hex ist PSPACE-vollständig, *Acta Informatica* **15**, 167–191.

312. M. Richardson [1953], Solutions of irreflexive relations, *Ann. of Math.* **58**, 573–590.

313. R.D. Ringeisen [1974], Isolation, a game on a graph, *Math. Mag.* **47**, 132–138.

314. R.L. Rivest, A.R. Meyer, D.J. Kleitman, K. Winklman and J. Spencer [1980], Coping with errors in binary search procedures, *J. Comput. System Sci.* **20**, 396–404.

315. E. Robertson and I. Munro [1978], NP-completeness, puzzles and games, *Utilitas Math.* **13**, 99-116.

316. A.G. Robinson and A.J. Goldman [1989], The set coincidence game: complexity, attainability, and symmetric strategies, *J. Comput. System Sci.* **39**, 376–387.

317. A.G. Robinson and A.J. Goldman [1990], On Ringeisen's isolation game, *Discrete Math.* **80**, 297–312.

318. A.G. Robinson and A.J. Goldman [≥1990], On Ringeisen's isolation game, II, *Discrete Math.*, in press.

319. A.G. Robinson and A.J. Goldman [1990], On the set coincidence game, *Discrete Math.* **84**, 261–283.

320. J.M. Robson [1983], The complexity of Go, Proc. Information Processing 83 (R.E.A. Mason, ed.), Elsevier, Amsterdam, pp. 413–417.

321. J.M. Robson [1984], Combinatorial games with exponential space complete decision problems, Proc. 11th Symp. Math. Foundations of Computer Science (M.P. Chytie and V. Koubek, eds.), Praha, Czechoslovakia, 1984, *Lecture Notes in Computer Science* **176**, 498–506, Springer, Berlin.

322. J.M. Robson [1984], N by N checkers is Exptime complete, *SIAM J. Comput.* **13**, 252–267.

323. J.M. Robson [1985], Alternation with restrictions on looping, *Inform. and Control* **67**, 2–11.

324. A.E. Roth [1978], A note concerning asymmetric games on graphs, *Naval Res. Logist. Quart.* **25**, 365–367.

325. A.E. Roth [1978], Two-person games on graphs, *J. Combin. Theory* (Ser. B) **24**, 238–241.

326. E.M. Rounds and S.S. Yau [1974], A winning strategy for SIM, *J. Recr. Math.* **7**, 193–202.

327. W.L. Ruzzo [1980], Tree-size bounded alternation, *J. Comput. Systems Sci.* **21** (1980), 218–235.

328. S. Sackson [1946], *A Gamut of Games*, Random House.

329. M. Saks and A. Wigderson [1986], Probabilistic Boolean decision trees and the complexity of evaluating game trees, 27th Annual Symp. Foundations of Computer Science (Toronto, Ont., Canada), IEEE Computer Soc., Washington, DC, pp. 29–38.

330. M. Sato [1972], Grundy functions and linear games, Publ. Res. Inst. Math. Sciences, Kyoto Univ., Vol. 7, 645–658.

331. W.L. Schaaf [1973, 1972, 1973, 1978], *A Bibliography of Recreational Mathematics* (four volumes), Nat'l. Council of Teachers of Mathematics, Reston, VA.

332. T.J. Schaefer [1976], Complexity of decision problems based on finite two-person perfect information games, 8th Annual ACM Symp. Theory of Computing (Hershey, PA, 1976), Assoc. Comput. Mach., New York, NY, pp. 41–49.

333. T.J. Schaefer [1978], On the complexity of some two-person perfect-information games, *J. Comput. System Sci.* **16**, 185–225.

334. G. Schmidt and T. Ströhlein [1985], On kernels of graphs and solutions of games: a synopsis based on relations and fixpoints, *SIAM J. Alg. Disc. Math.* **6**, 54–65.

335. G. Schrage [1985], A two-dimensional generalization of Grundy's game, *Fibonacci Quart.* **23**, 325–329.

336. H. Schubert [1953], *Mathematische Mussestunden*, Neubearbeitet von F. Fitting, Elfte Auflage, De Gruyter, Berlin.

337. F. Schuh, Spel van delers (The game of divisors) [1952], *Nieuw Tijdschrift voor Wiskunde* **39**, 299–304.

338. F. Schuh [1968], *The Master Book of Mathematical Recreations*, translated by F. Göbel, edited by T.H. O'Beirne, Dover, New York, NY.

339. B.L. Schwartz [1971], Some extensions of Nim, *Math. Mag.* **44**, 252–257.

340. A.J. Schwenk [1970], Take-away games, *Fibonacci Quart.* **8**, 225–234.

341. L.E. Shader [1978], Another strategy for SIM, *Math. Mag.* **51**, 60–64.

342. A.S. Shaki [1979], Algebraic solutions of partizan games with cycles, *Math. Proc. Camb. Phil. Soc.* **85**, 227–246.

343. A. Shamir, R.L. Rivest and L.M. Adleman [1981], Mental Poker, in: *The Mathematical Gardner* (D.A. Klarner, ed.), Wadsworth Internat., Belmont, CA, pp. 37–43.

344. W.L. Sibert and J.H. Conway [ ], Mathematical Kayles, submitted.

345. R. Silber [1976], A Fibonacci property of Wythoff pairs, *Fibonacci Quart.* **14**, 380–384.

346. R. Silber [1977], Wythoff's Nim and Fibonacci representations, *Fibonacci Quart.* **15**, 85–88.

347. R. Silver [1967], The group of automorphisms of the game of 3-dimensional ticktacktoe, *Amer. Math. Monthly* **74**, 247–254.

348. D.L. Silverman [1971], *Your Move*, McGraw-Hill.

349. D. Singmaster [1981], Almost all games are first person games, *Eureka* **41**, 33–37.

350. D. Singmaster [1982], Almost all partizan games are first person and almost all impartial games are maximal, *J. Combin. Inform. System Sci.* **7**, 270–274.

351. C.A.B. Smith [1966], Graphs and composite games, *J. Combin. Theory* **1**, 51-81. Reprinted in slightly modified form in: *A Seminar on Graph Theory* (F. Harary, ed.), Holt, Rinehart and Winston, New York, NY, 1967.

352. C.A.B. Smith [1968], Compound games with counters, *J. Recr. Math.* **1**, 67–77.

353. C.A.B. Smith [1971], Simple game theory and its applications, *Bull. Inst. Math. Appl.* **7**, 352–357.

354. D.E. Smith and C.C. Eaton [1911], Rithmomachia, the great medieval number game, *Amer. Math. Montly* **18**, 73–80.

355. J. Spencer [1977], Balancing games, *J. Combin. Theory* (Ser. B) **23**, 68–74.

356. J. Spencer [1984], Guess a number with lying, *Math. Mag.* **57**, 105–108.

357. J. Spencer [1986], Balancing vectors in the max norm, *Combinatorica* **6**, 55–65.

358. R. Sprague[1935–36], Über mathematische Kampfspiele, *Tôhoku Math. J.* **41**, 438–444.

359. R. Sprague [1937], Über zwei Abarten von Nim, *Tôhoku Math. J.* **43**, 351–359.

360. R. Sprague [1947], Bemerkungen über eine spezielle Abelsche Gruppe, *Math. Z.* **51**, 82–84.

361. R. Sprague [1961], *Unterhaltsame Mathematik*, Vieweg and Sohn, Braunschweig, Paperback reprint, translation by T.H. O'Beirne: *Recreations in Mathematics*, Blackie, 1963.

362. H. Steinhaus [1960], Definitions for a theory of games and pursuit, *Naval Res. Logist. Quart.* **7**, 105–108.

363. V.N. Stepanenko [1975], Grundy games under conditions of semidefiniteness, *Cybernetics* **11**, 167–172 (trans. of *Kibernetika* **11** (1975) 145–149).

364. D.L. Stock [1989], Merlin's magic square revisited, *Amer. Math. Monthly* **96**, 608–610.

365. L.J. Stockmeyer and A.K. Chandra [1979], Provably difficult combinatorial games, *SIAM J. Comput.* **8**, 151–174.

366. J.A. Storer [1983], On the complexity of chess, *J. Comput. System Sci.* **27**, 77–100.

367. Th. Ströhlein and L. Zagler [1977], Analyzing games by Boolean matrix iteration, *Discrete Math.* **19**, 183–193.

368. K. Sugihara and I. Suzuki [1989], Optimal algorithms for a pursuit-evasion problem in grids, *SIAM J. Disc. Math.* **1**, 126–143.

369. K. Sutner [1990], The σ-game and cellular automata, *Amer. Math. Monthly* **97**, 24–34.

370. G. Tardos [1988], Polynomial bound for a chip firing game on graphs, *SIAM J. Disc. Math.* **1**, 397–398.

371. M. Tarsi [1983], Optimal search on some game trees, *J. Assoc. Comput. Mach.* **30**, 389–396.

372. R. Telgársky [1987], Topological games: on the 50th anniversary of the Banach-Mazur game, *Rocky Mountain J. Math.* **17**, 227–276.

373. R. Tošić and S. Šćekić [1983], An analysis of some partizan graph games, Proc. 4th Yugoslav Seminar on Graph Theory, Novi Sad, pp. 311–319.

374. A.M. Turing, M.A. Bates, B.V. Bowden and C. Strachey[1953], Digital computers applied to games, in: *Faster Than Thought* (B.V. Bowden, ed.), Pitman, London, pp. 286–310.

375. R. Uehara and S. Iwata [1990], Generalized Hi-Q is NP-complete, *Trans. IEICE* **E73**, 270–273.

376. J. Úlehla [1980], A complete analysis of von Neumann's Hackendot, *Internat. J. Game Theory* **9**, 107–113.

377. J. van Leeuwen [1976], Having a Grundy-numbering is NP-complete, Report No. 207, Computer Science Dept., Pennsylvania State University, University Park, PA.

378. I.P. Varvak [1968], Games on the sum of graphs, *Cybernetics* **4**, 49–51 (trans. of *Kibernetika* **4** (1968) 63–66).

379. H. Venkateswaran and M. Tompa [1989], A new pebble game that characterizes parallel complexity classes, *SIAM J. Comput.* **18**, 533–549.

380. J. von Neumann [1928], Zur Theorie der Gesellschaftsspiele, *Math. Ann.* **100**, 295–320.

381. J. von Neumann and O. Morgenstern [1953], *Theory of Games and Economic Behaviour*, 3rd ed., Princeton University Press, Princeton, NJ.

382. W.A. Webb [1982], The length of the four-number game, *Fibonacci Quart.* **20**, 33–35.

383. C.P. Welter [1952], The advancing operation in a special abelian group, *Nederl. Akad. Wetensch. Proc.* (Ser. A) **55** = *Indag. Math.* **14**, 304–314.

384. C.P. Welter [1954], The theory of a class of games on a sequence of squares, in terms of the advancing operation in a special group, *Nederl. Akad. Wetensch. Proc.* (Ser. A) **57** = *Indag. Math.* **16**, 194–200.

385. M.J. Whinihan [1963], Fibonacci Nim, *Fibonacci Quart.* **1**(4), 9–12.

386. R. Wilber [1988], White pebbles help, *J. Comput. System Sci.* **36**, 108–124.

387. R.M. Wilson [1974], Graph puzzles, homotopy and the alternating group, *J. Combin. Theory* (Ser. B) **16**, 86–96.

388. W.A. Wythoff [1907], A modification of the game of Nim, *Nieuw Arch. Wisk.* **7**, 199–202.

389. A.M. Yaglom and I.M. Yaglom [1967], *Challenging Mathematical Problems with Elementary Solutions*, translated by J. McCawley, Jr., revised and edited by B. Gordon, Vol. II, Holden-Day, San Francisco.

390. Y. Yamasaki [1978], Theory of division games, Publ. Res. Inst. Math. Sciences, Kyoto Univ., Vol. 14, pp. 337–358.

391. Y. Yamasaki [1980], On misère Nim-type games, *J. Math. Soc. Japan* **32**, 461–475.

392. Y. Yamasaki [1981], The projectivity of Y-games, Publ. Res. Inst. Math. Sciences, Kyoto Univ., Vol. 17, pp. 245–248.

393. Y. Yamasaki [1981], Theory of Connexes I, Publ. Res. Inst. Math. Sciences, Kyoto Univ., Vol. 17, pp. 777–812.

394. Y. Yamasaki [1985], Theory of connexes II, Publ. Res. Inst. Math. Sciences, Kyoto Univ., Vol. 21, 403–410.

395. Y. Yamasaki [1989], *Combinatorial Games: Back and Front* (in Japanese), Springer Verlag, Tokyo.

396. Y. Yamasaki [≥1990], A difficulty in particular Shannon-like games, *Discrete Math.*, in press.

397. Y.K. Yu and R.B. Banerji [1982], Periodicity of Sprague-Grundy function in graphs with decomposable nodes, *Cybernetics and Systems: An Internat. J.* **13**, 299–310.

398. S. Zachos [1988], Probabilistic quantifiers and games, *J. Comput. System Sci.* **36**, 433–451.

399. E. Zermelo [1912], Über eine Anwendung der Mengenlehre auf die Theorie des Schachspiels, Proc. 5th Int. Cong. Math., Cambridge, Cambridge University Press, 1913, Vol. II, pp. 501–504.

400. W.S. Zwicker [1987], Playing games with games: the hypergame paradox, *Amer. Math. Monthly* **94**, 507–514.

Index